Miguel Ramos Pascual

The deuterium cycle (D-cycle) in fusion reactors

Miguel Ramos Pascual

The deuterium cycle (D-cycle) in fusion reactors

Analysis of fusion cross sections, reactivities and Lawson criterion with the SIMURGH code in tokamaks and stellarators

Noor Publishing

Imprint
Any brand names and product names mentioned in this book are subject to trademark, brand or patent protection and are trademarks or registered trademarks of their respective holders. The use of brand names, product names, common names, trade names, product descriptions etc. even without a particular marking in this work is in no way to be construed to mean that such names may be regarded as unrestricted in respect of trademark and brand protection legislation and could thus be used by anyone.

Cover image: www.ingimage.com

Publisher:
Noor Publishing
is a trademark of
Dodo Books Indian Ocean Ltd., member of the OmniScriptum S.R.L Publishing group
str. A.Russo 15, of. 61, Chisinau-2068, Republic of Moldova Europe
Printed at: see last page
ISBN: 978-620-3-86037-5

The deuterium cycle (D-cycle) in fusion reactors: an analysis of fusion cross sections, reactivities and Lawson criterion with the SIMURGH code in tokamaks and stellarators

Abstract

Fusion reactions in the deuterium cycle (D+D, D+T and D+^{3}He) are the main nucleus-nucleus interactions which occur in tokamaks and stellerators. In this paper several studies of fusion cross-sections and fusion reactivities $<\sigma v>$ have been reviewed and compared with experimental data from the last ENDF/B-VIII.0 cross-section library. Fusion reactivities have been estimated through a numerical integration of ENDF fusion cross-sections and compared with several formulations. Although relative differences between fusion cross-sections and reactivities are lower than 5 % between all cases and ENDF cross sections, there are considerable differences between resonance energies and partial widths (FWHM). In the D(d,n)^{3}He reaction, resonance energy is E_0=3821.4 keV for case 1, E_0=5012.7 keV for case 2, and E_0=931.3 keV for case 3. Furthermore, in the D(d,p)T reaction, resonance energy is E_0= 2798.2 keV for case 1, and E_0=931.3 keV for case 3, while case 2 does not present any resonance at energies lower than 8 MeV. Lawson criterion has been appplied to D+T fuel plasma and compared with several fusion reactors. As a conclusion, this shows that physical models for the astrophysical factor S(E) and bremstrahlung losses should be reviewed, in order to apply fusion to energy production in safety conditions.

Keywords: Fusion reaction, fusion cross-section, fusion reactivity, astrophysical S-factor, Gamow-Sommerfeld factor, SLBW, Lawson criterion

1. Introduction

Fusion reactions in the deuterium cycle (D+D, D+T and D+^{3}He) are the main nucleus-nucleus interactions which occur in fusion reactors [Smith and Cowley 2010, Donné 2018, Zohm 2019]. These reactions are the limiting case between the Woods-Saxon potential field at nuclear distances and the Coulomb electrostatic potential (scattering) at longer distances. In order to overpass the Coulomb electrostatic barrier and achieve nuclear distances to allow fusion reactions, ionized plasma is confined by a magnetic field inside tokamaks and stellerators [Knoepfel 1978, Woods 2006]

Fusion cross sections are applied to characterise functions such as the average plasma temperature, fusion reaction rates and fusion reaction probabilities. Fusion cross-sections have been previously analysed from solar measurements, specially astrophysical S-factor functions [Adelberger 1998, 2011] During decades, fusion cross-sections have been estimated with beam-target experiments on accelerators and interpolated through several parametric equations to quantum mechanical models (R-matrix) [Bosch and Hale 1992, Li et al 2006, 2008, NRL 2019].

Fusion reactivities ($<\sigma v>$) have been estimated in the last decades through several approaches, considering Maxwell-Boltzmann distribution of ion velocities and fitting cross-section data. Several formulation and codes that calculate fusion reactivities from the D-D and D-T reactions are analysed in Hively [1983], BUCKY [McFarlane et al 2002], DRACO [Heltemes et al 2005], T2 LANL, Langenbrunner and Booker [2017].

The SIMURGH code has been developed to estimate mainly fusion cross-section, fusion reactivities $<\sigma v>$, geometric, Gamow-Sommerfeld, S-factors and Single-Level-Breit-Wigner (SLBW) resonance parameters in the D-cycle. Fusion cross-sections equations estimated through different expression have been compared with ENDF/B-VIII.0 cross-section libraries: Duane's 5-parameter formula (case 1), Bosch-Hale 9-parameter formula though a Padé expansion (case 2) and a combined geometric factor based on wavenumber and a complex phase-shift S-factor (case 3).

2. The D-cycle

The deuterium cycle of fusion is composed of several interdependent nuclear reactions [Atzeni and Meyer-ter-Vehn 2001, Ball 2019]:

$$
\begin{array}{llll}
{}^{2}_{1}D + {}^{2}_{1}D & \rightarrow {}^{3}_{1}T \text{ (1.01 MeV)} & + & {}^{1}_{1}H \text{ (3.02 MeV)} \\
{}^{2}_{1}D + {}^{2}_{1}D & \rightarrow {}^{3}_{2}He \text{ (0.82 MeV)} & + & {}^{1}_{0}n \text{ (2.45 MeV)} \\
{}^{2}_{1}D + {}^{3}_{1}T & \rightarrow {}^{4}_{2}He \text{ (3.52 MeV)} & + & {}^{1}_{0}n \text{ (14.07 MeV)} \\
{}^{2}_{1}D + {}^{3}_{2}He & \rightarrow {}^{4}_{2}He \text{ (3.67 MeV)} & + & {}^{1}_{1}H \text{ (14.68 MeV)}
\end{array}
$$

which can be summarized in

$$
\begin{array}{ll}
6\,{}^{2}_{1}D \rightarrow & {}^{4}_{2}He \text{ (3.52 MeV)} + {}^{1}_{1}H \text{ (3.02 MeV)} + {}^{1}_{0}n \text{ (14.07 MeV)} \\
& {}^{4}_{2}He \text{ (3.67 MeV)} + {}^{1}_{1}H \text{ (14.68 MeV)} + {}^{1}_{0}n \text{ (2.45 MeV)} + 1.83 \text{ MeV}
\end{array}
$$

Furthermore, protons produced by fusion reactions can also react with low probability as

$$
\begin{array}{lll}
{}^{1}_{1}H + {}^{1}_{1}H & \rightarrow {}^{2}_{1}D + e^{+} + \upsilon + 0.93 \text{ MeV} & (99.76\ \%) \\
{}^{1}_{1}H + {}^{1}_{1}H + e^{-} & \rightarrow {}^{2}_{1}D + \upsilon + 1.95 \text{ MeV} & (0.24\ \%)
\end{array}
$$

Other fusion reactions of the deuterium cycle with a low branching ratio are [Robouch et al 1993, Bystritsky et al 1998]

$$
{}^{2}_{1}D + {}^{2}_{1}D \rightarrow {}^{4}_{2}He \text{ (0.08 MeV)} + \gamma_0 \text{(23.77 MeV)}
$$

and with tritium as target nuclei

$$
\begin{array}{lll}
{}^{2}_{1}D + {}^{3}_{1}T \rightarrow & {}^{4}_{2}He \text{ (3.52 MeV)} & + {}^{1}_{0}n \text{ (14.07 MeV)} \\
{}^{2}_{1}D + {}^{3}_{1}T \rightarrow & {}^{5}_{2}He^{*} \text{ (0.05 MeV)} & + \gamma_0 \text{(16.65 MeV)} \\
{}^{2}_{1}D + {}^{3}_{1}T \rightarrow & {}^{5}_{2}He^{*} \text{ (3.2 MeV)} & + \gamma_1 \text{(13.50 MeV)}
\end{array}
$$

There are other fusion reactions in which the reaction heat is positive, which have not been considered, such as

$$
\begin{array}{lll}
{}^{2}_{1}D + {}^{3}_{1}T & \rightarrow {}^{3}_{2}He & + 2\,{}^{1}_{0}n \\
{}^{2}_{1}D + {}^{3}_{2}He & \rightarrow {}^{3}_{2}He + {}^{1}_{1}H & + {}^{1}_{0}n
\end{array}
$$

Table 1 summarizes the main characteristics of the D-cycle fusion: branching ratios (BR), reaction heat (Q) and emergent particle/gamma energies.

3. Fusion cross sections in the D-cycle

Fusion cross-sections are characterised by three energy dependent functions: geometric, Gamow-Sommerfeld and astrophysical S-factor. These factors retain the physical characteristics of fusion reaction: a Coulomb electrostatic potential barrier penetration, in a form of an exponential attenuation, G(E), and a strong force interaction after tunneling effect in the nuclear potential well (Woods-Saxon), at the boundary conditions, in a form of an astrophysical S(E) factor or resonance function [Miley et al 1974, Sadler and van Belle 1987, Li 2002, Chen 2016].

Fusion cross section is expressed in a simplified form as

$$\sigma_f(E) = \phi GS$$

where ϕ is the geometric factor, a term inversely proportional with incident particle energy, S is a strong-energy correlated parameter S-factor, called also astrophysical or S-factor, which depends on the boundary conditions between nuclear well and Coulomb electrostatic field and G(E), a tunneling or penetration factor, called Gamow-Sommerfeld factor. In particular, the astrophysical S-factor is a resonance function, based on the classical formula of Breit-Wigner of single-level resonances [Breit and Wigner 1936].

Fusion cross-section of light nuclei were firstly estimated by the Naval Research Laboratory, who published a 5 parameter formula for D+D, D+T and D+^{3}He light nuclei fusion reactions, also known as Duane 5-parameter formula, based on accelerator beam-target experiments [NRL 2019].

Further studies, such as those carried out by Bosch and Hale in 1992, included a 9 parameter formula for Gamow and S-factor, based on the R-matrix of thousand experiments, and an inverse energy dependent geometric factor. However, there are some discrepances between other fusion cross-sections. For example, in the case of the D(d,p)T fusion reaction, the S-factor has no resonance peak.

In the last decades, fusion cross-sections have been modeled with a complex phase-shift of the incident wavefunction at the boundaries of Coulomb electrostatic potential and Woods-Saxon potential for the nuclear well, including SLBW resonance peaks for all target nuclei (D, T or ^{3}He) [Li et al 2006,2008, Singh et al 2019].

4. Geometric, Gamow-Sommerfeld and S-factor

4.1. Geometric factor (ϕ)

In most of the fusion cross-section expressions, geometric factor is expressed as $\phi = 1/E$. In the last decades, geometric factor has been expressed as the area subtended by a de Broglie wavelength $\lambda\!\!\!^{-}$ or wavenumber *k*, which depends on energy [Li et al 2006, 2008]

$$\phi(E) = \pi \lambda\!\!\!^{-2} = \frac{\pi}{k^2}$$

with

$$k = \sqrt{2\mu E}/\hbar$$

The geometric factor $\phi(E)$ can also be expressed with two independent terms,

$$\phi(E) = \pi \left(\frac{\hbar^2}{2\mu} \right) \frac{1}{E}$$

The first term is a conversion factor which is included in some expressions in the Gamow-Sommerfeld or in the S-factor, with units of *keV.b* in the SI.

4.2. Gamow-Sommerfeld factor

4.2.1 Introduction

The Gamow-Sommerfeld factor *G(E)* is a penetration factor of the Coulomb barrier by the incident particle and characterize the probability of incident/target nuclei of reaching a fusion reaction. It is expressed with an exponential attenuation function as [Brennan and Coyne 1964, Humblet et al 1987]

$$G(E) = \frac{1}{\chi^2} = \exp(-2\pi\eta_l)$$

with η_l the Sommerfeld parameter, evaluated as

$$\eta_l = \frac{1}{\pi} \int_{r_n}^{r_{tp}} |k_l(r)| dr$$

with k_l the wave number inside the nuclear well

$$k_l(r) = \left[\frac{2\mu}{\hbar^2} (E - U(r)) \right]^{1/2}$$

where *U(r)* is the potential field, r_n is the classical distance of closest approach (nuclear square-well radius) and r_{tp} is the classical turning point, that is, the distance in which the energy of the incident particle energy is $E = U(r_{tp})$ [Lee and Jung 2017]

4.2.2 Case 1: Duane parametrization

Duane parametrization included an interpolation of the Gamow factor as the Mott's form of the Coulomb barrier penetrability [Mott and Massey 1965, NRL 2019]

$$G(E) = \frac{1}{\exp\left[A1/\sqrt{E}\right] - 1}$$

4.2.3 Case 2

Let us consider that the potential field *U(r)* can be approximated as

$$U(r) = U_C(r) + \frac{\hbar^2 l(l+1)}{2\mu r^2}$$

where U_C is the Coulomb potential and the second term is the spin-orbit potential, with *l* the magnetic quantum number.

The Coulomb electrostatic potential is expressed as

$$U_C(r) = k_e \frac{Z_1 Z_2 e^2}{r} \qquad r \geq r_n$$

with $k_e = 1/4\pi\varepsilon_0$ the Coulomb electrostatic constant and r_n is estimated as

$$r_n = r_0\left[A_1^{1/3} + A_2^{1/3}\right]$$

with r_0 the specific radius per nucleon of fusion reaction and the classical turning point r_{tp} as

$$r_{tp} = k_e Z_1 Z_2 e^2 / E$$

The centrifugal term is zero at the boundaries of the nuclear potential well (*l*=0), then the integral is calculated as [Atzeni 2004]

$$\eta_0 = \frac{1}{\pi^2}\frac{B_g}{\sqrt{E}}\left(\arccos\left(\sqrt{\frac{E}{U_0}}\right) - \left[\sqrt{\frac{E}{U_0}}\sqrt{\left(1 - \frac{E}{U_0}\right)}\right]\right)$$

where

$$U_0 = U(r_n) = E\frac{r_{tp}}{r_n}$$

and

$$B_g = \varepsilon_g^{1/2} = 2\pi\, k_e Z_1 Z_2 e^2 \sqrt{\mu / 2\hbar^2}$$

is the Gamow constant.

The η_0 can be also expressed in term of the lengths r_n and r_{tp} as

$$\eta_0 = \frac{1}{\pi^2}\frac{B_g}{\sqrt{E}}\left(\arccos\left(\sqrt{\frac{r_n}{r_{tp}}}\right) - \left[\sqrt{\frac{r_n}{r_{tp}}\left(1-\frac{r_n}{r_{tp}}\right)}\right]\right)$$

The arccosine can be expanded in powers of $(E/U_0)^{1/2}$, thus

$$\eta_0 = \frac{1}{\pi^2}\frac{B_g}{\sqrt{E}}\left(\left[\frac{\pi}{2} - \left(\left(\frac{E}{U_0}\right)^{1/2} + \frac{1}{6}\left(\frac{E}{U_0}\right)^{3/2} + \frac{3}{40}\left(\frac{E}{U_0}\right)^{5/2} + \ldots\right)\right] - \left[\sqrt{\frac{E}{U_0}}\sqrt{\left(1-\frac{E}{U_0}\right)}\right]\right)$$

In the low energy limit, with $E << U_0$, η_0 reduces to the Sommerfeld parameter

$$\eta_0 = \frac{1}{2\pi}\frac{B_g}{\sqrt{E}}$$

and the Gamow factor to

$$G(E) = \exp\left(-B_g/\sqrt{E}\right)$$

The Sommerfeld parameter can be expressed also as $\eta_0 = \alpha Z_1 Z_2 (c/v)$, with $\alpha = e^2/(4\pi\varepsilon_0 \hbar c)$ the fine structure constant, and v the relative velocity of the incident particle, related with the kinetic energy as, $E = \mu v^2/2$.

4.2.4 Case 3

In some approximations, the Gamow factor is expressed in a similar way to the Sommerfeld expression as [Yoon and Wong 2000]

$$G(E) = \frac{2\pi}{\exp(2\pi\eta_0) - 1}$$

which is equivalent to [Li et al 2006,2008]

$$G(E) = \frac{2\pi}{\exp(2\pi/ka_c) - 1}$$

with a_c the Coulomb unit of length

$$a_c = \frac{\hbar^2}{k_e Z_1 Z_2 \mu e^2}$$

Other expressions have included relativistic considerations about Gamow-Sommefeld factors, which will be considered in future research [Arbuzov and Kopylova 2012]

4.3. S-factor

4.3.1 Introduction

The S-factor is a function which characterizes the width and energies of the resonance peaks of a nuclear reaction. The Breit-Wigner formula was firstly applied to the non-relativistic scattering of slow neutrons with dependence only on resonance energy position and width [Breit and Wigner 1936]

In the case of a single level Breit-Wigner resonance E_0, the S-factor is

$$S(E) = \frac{\Gamma_i \Gamma_f}{(E - E_0)^2 + (\Gamma/2)^2}$$

where E_0 is the energy of the single-level resonance, Γ_i and Γ_f are the partial widths of the entrance and exit channels of the fusion reaction and $\Gamma = \sum_j \Gamma_j$ is the total decay width, equal to the width of the resonance peak at half height [Takano and Ishiguro 1976, Blatt and Weisskopf 1979, Belov et al 2019].

4.3.2 Case 1

The S-factor was interpolated by Duane through a [1/2] order Padé approximation with a constant term as [Duane 1972, Miley et al 1974, NRL 2019]

$$S(E) = \frac{A_2}{(A_4 - A_3 E)^2 + 1} + A_5$$

where the single-level Breit-Wigner (SLBW) resonance is at $E_0 = A_4 / A_3$ and the total resonance width is $\Gamma = 2 / A_3$ (see table 2) In the case of D+D fusion reactions, in which A_5=0, the S-factor equals the classical Breit-Wigner resonance formula.

4.3.3 Case 2

Further approaches increased the order of the interpolation, through a [4/4] order Padé approximation as [Bosch and Hale 1992]

$$S(E) = \frac{A_1 + E(A_2 + E(A_3 + E(A_4 + EA_5)))}{1 + E(B_1 + E(B_2 + E(B_3 + EB_4)))}$$

Table 3 presents the coefficients for the [4/4] order Padé approximation. In the case of fusion reactions $T(d,n)^4He$ and $3He(d,p)^4He$, the S-factor was interpolated in two different energy regions: [0.5-550 keV] and [550-4700 keV] for $T(d,n)^4He$ and [0.3-900 keV] and [900-4800 keV] for $3He(d,p)^4He$.

4.3.4 Case 3

Let us consider that the Woods-Saxon potential in the nuclear well is modeled as a square-well complex potential as

$$U_S(r) = U_r + iU_i$$

where U_r and U_i are the real and imaginary part of the potential [Mohr 1957].

The complex phase shift $\omega(\delta_0)$ of the incident wavefunction is

$$\omega(\delta_0) = \omega_r + i\omega_i = W / \chi^2 = (W_r + iW_i) / \chi^2 = \cot\delta_0 / \chi^2$$

In this case, the dimensionless S-factor is defined through the function $\omega(\delta_0)$ as [Zhong-Li et al 2004, Li et al 2008, Jing et al 2009, Singh et al 2019]

$$S(E) = \frac{-4\omega_i}{\omega_r^2 + (\omega_i - (1/\chi^2))^2}$$

Let us define the complex function

$$\gamma = \gamma_r + i\gamma_i = (K_r + iK_i)r_n = Kr_n$$

where r_n is the radius of the nuclear well, estimated as

$$r_n = r_0\left[A_1^{1/3} + A_2^{1/3}\right]$$

with r_0 the specific radius per nucleon of the nuclear well, which depends on target/incident nuclei.

Therefore, the wavenumber K is

$$K^2 = (K_r + iK_i)^2 = a + ib$$

where

$$a = \frac{2\mu}{\hbar^2}(E - U_r)$$
$$b = \frac{2\mu}{\hbar^2}(-U_i)$$

Equaling real and imaginary parts,

$$K_r^2 - K_i^2 = a$$
$$2K_rK_i = b$$

Hence,

$$K_r = \left(\frac{a + \sqrt{a^2 + b^2}}{2}\right)^{1/2}$$
$$K_i = \frac{b}{2K_r}$$

The boundary conditions of the wavefunction Ψ in the nuclear well as

$$\psi(r_n) = r_n \frac{[\sin(kr)]'}{\sin(kr)}\bigg|_{r=r_n} = Kr_n \frac{\cos(Kr_n)}{\sin(Kr_n)} = \gamma \cot(\gamma)$$

and in the Coulomb field as [Landau and Lifshitz 1987]

$$\psi(r_n) = \frac{r_n}{a_c}\left\{\frac{1}{\chi^2}\cot(\delta_0) + 2H\right\}$$

Therefore, equaling both conditions

$$\gamma \cot(\gamma) = \frac{r_n}{a_c}\{\omega + 2H\}$$

then,

$$\omega = \cot(\delta_0)/\chi^2 = \frac{a_c}{r_n}\gamma\cot(\gamma) - 2H$$

The real and imaginary part of the phase function ω are obtained through

$$
\begin{aligned}
\omega_i &= \mathrm{Im}\left[\frac{a_c}{r_n}\gamma\cot(\gamma)\right] \\
&= \frac{a_c}{r_n}\mathrm{Im}\left[(\gamma_r + i\gamma_i)\frac{\cos(\gamma_r + i\gamma_i)}{\sin(\gamma_r + i\gamma_i)}\right] \\
&= \frac{a_c}{r_n}\mathrm{Im}\left[(\gamma_r + i\gamma_i)\frac{\cos\gamma_r\cosh\gamma_i - i\sin\gamma_r\sinh\gamma_i}{\sin\gamma_r\cosh\gamma_i + i\cos\gamma_r\sinh\gamma_i}\right] \\
&= \frac{a_c}{r_n}\mathrm{Im}\left[(\gamma_r + i\gamma_i)\left(\frac{\cos\gamma_r\cosh\gamma_i - i\sin\gamma_r\sinh\gamma_i}{\sin\gamma_r\cosh\gamma_i + i\cos\gamma_r\sinh\gamma_i}\right)\left(\frac{\sin\gamma_r\cosh\gamma_i - i\cos\gamma_r\sinh\gamma_i}{\sin\gamma_r\cosh\gamma_i - i\cos\gamma_r\sinh\gamma_i}\right)\right] \\
&= \frac{a_c}{r_n}\mathrm{Im}\left[(\gamma_r + i\gamma_i)\left(\frac{\sin\gamma_r\cos\gamma_r(\cosh^2\gamma_i - \sinh^2\gamma_i) - i\sinh\gamma_i\cosh\gamma_i(\sin^2\gamma_r + \cos^2\gamma_r)}{\sin^2\gamma_r\cosh^2\gamma_i + \cos^2\gamma_r\sinh^2\gamma_i}\right)\right] \\
&= \frac{a_c}{r_n}\mathrm{Im}\left[\frac{\begin{array}{l}\{\gamma_r\sin\gamma_r\cos\gamma_r(\cosh^2\gamma_i - \sinh^2\gamma_i) + \gamma_i\sinh\gamma_i\cosh\gamma_i(\sin^2\gamma_r + \cos^2\gamma_r)\} \\ + i\{\gamma_i\sin\gamma_r\cos\gamma_r(\cosh^2\gamma_i - \sinh^2\gamma_i) - \gamma_r\sinh\gamma_i\cosh\gamma_i(\sin^2\gamma_r + \cos^2\gamma_r)\}\end{array}}{\sin^2\gamma_r\cosh^2\gamma_i + \cos^2\gamma_r\sinh^2\gamma_i}\right] \\
&= \frac{a_c}{r_n}\left(\frac{\gamma_i\sin\gamma_r\cos\gamma_r(\cosh^2\gamma_i - \sinh^2\gamma_i) - \gamma_r\sinh\gamma_i\cosh\gamma_i(\sin^2\gamma_r + \cos^2\gamma_r)}{\sin^2\gamma_r\cosh^2\gamma_i + \cos^2\gamma_r\sinh^2\gamma_i}\right) \\
&= \frac{a_c}{2r_n}\left(\frac{\gamma_i\sin 2\gamma_r - \gamma_r\sinh 2\gamma_i}{\sin^2\gamma_r\cosh^2\gamma_i + \cos^2\gamma_r\sinh^2\gamma_i}\right) \\
&= \frac{a_c}{2r_n}\left(\frac{\gamma_i\sin 2\gamma_r - \gamma_r\sinh 2\gamma_i}{\sin^2\gamma_r(1 + \sinh^2\gamma_i) + (1 - \sin^2\gamma_r)\sinh^2\gamma_i}\right) \\
&= \frac{a_c}{2r_n}\left(\frac{\gamma_i\sin 2\gamma_r - \gamma_r\sinh 2\gamma_i}{\sin^2\gamma_r + \sinh^2\gamma_i}\right)
\end{aligned}
$$

and

$$\omega_r = \mathrm{Re}\left[\frac{a_c}{r_n}\gamma\cot(\gamma) - 2H\right]$$

$$= \frac{a_c}{r_n}\mathrm{Re}\left[\frac{\begin{matrix}\{\gamma_r \sin\gamma_r \cos\gamma_r(\cosh^2\gamma_i - \sinh^2\gamma_i) + \gamma_i \sinh\gamma_i \cosh\gamma_i(\sin^2\gamma_r + \cos^2\gamma_r)\} \\ +i\{\gamma_i \sin\gamma_r \cos\gamma_r(\cosh^2\gamma_i - \sinh^2\gamma_i) - \gamma_r \sinh\gamma_i \cosh\gamma_i(\sin^2\gamma_r + \cos^2\gamma_r)\}\end{matrix}}{\sin^2\gamma_r \cosh^2\gamma_i + \cos^2\gamma_r \sinh^2\gamma_i}\right] - 2H$$

$$= \frac{a_c}{r_n}\left(\frac{\gamma_r \sin\gamma_r \cos\gamma_r(\cosh^2\gamma_i - \sinh^2\gamma_i) + \gamma_i \sinh\gamma_i \cosh\gamma_i(\sin^2\gamma_r + \cos^2\gamma_r)}{\sin^2\gamma_r \cosh^2\gamma_i + \cos^2\gamma_r \sinh^2\gamma_i}\right) - 2H$$

$$= \frac{a_c}{2r_n}\left(\frac{\gamma_r \sin 2\gamma_r + \gamma_i \sinh 2\gamma_i}{\sin^2\gamma_r(1 + \sinh^2\gamma_i) + (1 - \sin^2\gamma_r)\sinh^2\gamma_i}\right) - 2H$$

$$= \frac{a_c}{2r_n}\left(\frac{\gamma_r \sin 2\gamma_r + \gamma_i \sinh 2\gamma_i}{\sin^2\gamma_r + \sinh^2\gamma_i}\right) - 2H$$

Hence,

$$\omega_i = W_i / \chi^2 = \frac{a_c}{2r_n}\left\{\frac{\gamma_i \sin(2\gamma_r) - \gamma_r \sinh(2\gamma_i)}{\sin^2(\gamma_r) + \sinh^2(\gamma_i)}\right\}$$

$$\omega_r = W_r / \chi^2 = \frac{a_c}{2r_n}\left\{\frac{\gamma_r \sin(2\gamma_r) + \gamma_i \sinh(2\gamma_i)}{\sin^2\gamma_r + \sinh^2\gamma_i}\right\} - 2H$$

The function H can be expressed as

$$H = \left[\ln(2ka_c) + \mathrm{Re}\left(\frac{\Gamma'(-1/ka_c)}{\Gamma(-1/ka_c)}\right) + 2A\right]$$

or as

$$H = [\ln(2ka_c) + y(ka_c) + 2A]$$

with A the Euler constant and $y(ka_c)$ related to the logarithmic derivative of the Γ function given by

$$y(x) = \frac{\Gamma'(-x^{-1})}{\Gamma(-x^{-1})} = \frac{1}{x^2}\sum_{n=1}^{\infty}\frac{1}{n(n^2 + x^{-2})} - A + \ln(x)$$

which can be approximated to

$$y(ka_c) = \frac{1}{12}(ka_c)^2$$

Table 7 presents the nuclear potential well parameters and average atomic radius for the D+D, D+T and D+^{3}He fusion reactions. Figure 1 shows the nuclear Woods-Saxon and Coulomb electrostatic potentials for fusion reactions, whereas figures 2 and 3 presents the complex real and imaginary part of the phase function ω. From the figures, the energies at which ω_r=0, corresponding with the SLBW resonances, are 1039.58 keV, 51.97 keV and 218.84 keV for D+D, D+T and D+^{3}He, respectively.

5. Fusion cross-sections

5.1 ENDF library

ENDF/B-VIII.0 (Evaluated Nuclear Data File) is the library containing the last revision of experimental fusion cross-section reactions from Brookhaven National Laboratory, based on accelerator beam-target experiments with incident particle (proton, deuteron, triton and helium-3) into several targets (H, D ,T, Helium) [Brown et al 2018]

ENDF/B-VIII.0 libraries are available in the EXFOR database (Experimental Nuclear Reaction Data), which contains an extensive compilation of experimental nuclear reaction data [Otuka 2014].

Table 5 summarizes all the fusion cross section cases that have been used, including geometric, Gamow and S-factors. These expressions have been compared with ENDF/B-VIII.0 cross sections data for fusion reactions of the D-cycle: D(d,p)T, D(d,n)3He, T(d,n)^{4}He and 3He(d,p)^{4}He.

Plasma temperature has been estimated as

$$T(K) = E(MeV)\left[\frac{e}{k_b}\right]$$

where E is the kinetic energy of the incident ions, e the electron charge and k_b=1.3807 x 10^{-23} J K^{-1} the Boltzmann constant.

5.2 Case 1

The fusion cross-section $\sigma_f(E)$ was firstly characterized by Duane's through a 5 parameter expression as

$$\sigma_f(E) = \frac{1}{E}\left(\frac{1}{\exp\left(A_1/\sqrt{E}\right)-1}\right)\left(\frac{A_2}{\left(A_3E-A_4\right)^2+1}+A_5\right)$$

which assumed the geometric factor as a term dependent only of energy, Gamow factor through Mott's form of the Coulomb barrier penetrability and the astrophysical S-factor as a [2/2] order Padé approximation of the classical Breit-Wigner resonance formula.

5.3 Case 2

In this case, the S-factor is estimated by a [4/4] Padé approximation and Gamow factor by its Gamow constant,

$$\sigma_f(E) = \frac{1}{E}\left(\frac{1}{\exp\left[B_G/\sqrt{E}\right]}\right)\left[\frac{A_1+E(A_2+E(A_3+E(A_4+EA_5)))}{1+E(B_1+E(B_2+E(B_3+EB_4)))}\right]$$

5.4 Case 3

In the last case, fusion cross-section is calculated as

$$\sigma_f(E) = \left(\frac{1}{\beta}\frac{2\mu}{\pi\hbar^2}\right)\left[\frac{\pi}{k^2}\right]\left(\frac{2\pi}{\exp[2\pi / ka_c]-1}\right)\left(\frac{-4\omega_i}{\omega_r^2 + \left[\omega_i - \left(1/\chi^2\right)\right]^2}\right)$$

where $1/\chi^2$ is the Gamow factor and β is a factor which depends on the target nuclei.
6. Fusion reactivities and power density

Fusion reactivities in thermal equilibrium at temperature *T* is defined as

$$<\sigma v>(T) = \int\int f(\vec{v}_i) f(\vec{v}_j) \sigma(v) v \, d\vec{v}_i d\vec{v}_j$$

where $v = \left|\vec{v}_i - \vec{v}_j\right|$ is the relative velocity between ions *i* and *j*, and *p* is the Maxwell-Boltzmann distribution of the of ion velocities, that is

$$p(v)\,dv = \left(\frac{2}{\pi}\right)^{1/2}\left(\frac{v^2}{a^3}\right)\exp\left[-v^2/2a^2\right]dv$$

where $a = \sqrt{k_b T / \mu}$, with *T* the plasma temperature, and *m* is the mass of the ions *i* and *j*,

$$\mu = m_i m_j / \left(m_i + m_j\right)$$

Hence, considering the normalised probability density function,

$$<\sigma v>(v,T) = \frac{\int_0^\infty v\left(\frac{2}{\pi}\right)^{1/2} v^2 \exp\left[-v^2/2\sqrt{kT/\mu}\right]\sigma(v,T)\,dv}{\int_0^\infty \left(\frac{2}{\pi}\right)^{1/2} v^2 \exp\left[-v^2/2\sqrt{kT/\mu}\right]dv}$$

which can be simplified to

$$<\sigma v>(v,T) = \frac{\int_0^\infty v^3 \exp\left[-v^2/2\sqrt{kT/\mu}\right]\sigma(v,T)\,dv}{\int_0^\infty v^2 \exp\left[-v^2/2\sqrt{kT/\mu}\right]dv}$$

Then, considering the reduction formula for the integral,

$$<\sigma v>(v,T) = \left(\sqrt{\frac{\pi}{8}k_b T/\mu}\right)^{-1}\int_0^\infty v^3 \exp\left[-v^2/2\sqrt{k_b T/\mu}\right]\sigma(v,T)\,dv$$

Figure 4 presents the Maxwell-Boltzmann distribution of speeds for the D+D reaction at different plasma temperatures from 1 keV to 1000 keV.

Fusion reactivities have been previously published in several studies and some software codes have been developed to this purpose [Langenbrunner and Brooker 2017], such as the Duane's parametrization in the NRL formulary, Hively-Kozlov formulation [Hively83], Peres formulation

[Peres79, Bosch and Hale 92], BUCKY [MacFarlane et al 2002], DRACO [Heltemes et al 2005] and UNC [UNC] Most of these formulation are polynomial fittings as a function of thermal temperature *T*, such as the Hively-Kozlov formulation for D+D, that is

$$<\sigma v>= A_1\left(1+A_3T^{3/4}\right)T^{-2/3}\exp\left(-\frac{A_2}{T^{1/3}}\right)$$

and for D+T and D+^{3}He

$$<\sigma v>= A_1\left(1+A_3T^{3/4}\right)T^{-2/3}\exp\left(-\frac{A_2}{T^{1/3}}\right)\left(\frac{1}{\left(1+A_4T^{3.25}\right)^{1/2}}\right)$$

An approach to fusion reactivity that is applied in many analysis is [Peres 1979, Bosch and Hale 1992]

$$<\sigma v>= C_1\theta\sqrt{\xi/\mu c^2T^3}\,e^{-3\xi}$$

$$\theta = T\left[1-\frac{T(C_2+T(C_4+TC_6))}{1+T(C_3+T(C_5+TC_7))}\right]^{-1}$$

$$\xi=\left(B_g^2/4\theta\right)^{1/3}$$

where B_g is the Gamow constant.

The power density in a plasma volume is estimated as

$$P=\frac{n_1n_2}{4}Q_{12}<\sigma v>$$

where n_1 and n_2 are the ion density and Q_{12} is the energy released per fusion reaction.

Fusion reactivities coefficients are described in detail in tables 6, 7 and 8 for the different models.

7. Lawson criterion

Lawson criterion states the energy balance between energy production by fusion reaction in a plasma and energy losses, mainly by thermal and bremstrahlung radiation of reaction products (i.e. alpha particle in D+T fuel plasma). In case of energy production, assuming that particle temperatures are equal in the plasma, Lawson criterion requieres that fusion heating exceeds losses [Lawson 1955],

$$n_1n_2\langle\sigma v\rangle E_f\eta \geq \frac{3}{2}\frac{(n_1+n_2)}{\tau_E}T + C_bn_1n_2Z^2T^{1/2}$$

and considering that the ion densities of deuterium and tritium are equal $n_1=n_2$, is

$$\frac{n^2}{4}\langle\sigma v\rangle E_f\eta \geq \frac{3nT}{\tau_E}+C_bn^2Z^2T^{1/2}$$

where T is the plasma temperature in keV, E_f is the energy of the fission products (alpha particle in D+T reaction) η is the efficiency conversion of fusion energy, τ_E is the energy confinement time in the plasma and C_b is the bremstrahlung radiation coefficient, which in case of a D+T plasma fuel is $C_b=5.34\times10^{-37}\,Wm^3keV^{-1/2}$ for the alpha particles in vacuum, and Z=1 in a hydrogenic plasma (deuterium and tritium ions) [Kikuchi et al 2012, Wurzel and Hsu 2021]

Rearranging terms and multiplying by T, we arrive to the triple product that defines the Lawson criterion,

$$n\tau_E T \geq \frac{12T^2}{\langle\sigma v\rangle E_f \eta - 4C_b T^{1/2}}$$

The efficiency η can be estimated with the expression

$$\eta = f_c + \frac{f_a}{Q_{fuel}}$$

where $f_c = 1$ is the energy fraction of fusion products in charged particles, $f_a = E_f / E_\alpha = 4.96$ is the energy ratio between fusion energy and energy of the alpha particle and Q_{fuel} is the fuel gain, the ratio of fusion power to absorbed heating power.

The ignition condition is achieved when $Q_{fuel} \to \infty$, that is when $\eta = f_c$.

The break-even condition is achieved when $Q_{fuel} = 1$, that is, $\eta = f_c + f_a$.

9. Results and discussion

Figures 5 to 13 show Gamow-Sommerfeld factors and astrophysical S-factors for different fusion reactions. Significant differences have been observed between Gamow and S-factors, in particular with SLBW resonance energy values and Γ (FWHM).

There are considerable differences between case 1 and the other cases considering resonance peak energy for T(d,n)4He and 3He(d,p)4He reactions. In case 1, energy resonances are placed at 78.7 MeV and 325.9 MeV, respectively, while for cases 2/3 are placed at 49.1/42 MeV and 215.3/207.8 MeV, as showed in table 9.

In case 2, the S-factor of the D(d,p)T reaction does not include the Breit-Wigner resonance peak as cases 1 and 3 in 2798.2 MeV and 931.3 keV, respectively. In addition, case 1 includes a single resonance at 2798.16 keV for reaction D(d,p)T and 3821.43 keV for reaction D(d,n)^{3}He, respectively, whereas case 2 has a resonance at 5012.7 keV for reaction D(d,n)^{3}He and case 3 has a single resonance in 931.3 keV for both reactions. As a notation, this energy value is equivalent to the conversion factor *keV* and *uma*, although no physical correlation has been observed between them.

Furthermore, D+D fusion reaction for case 3 is independent of reaction products (see fig. 11), while other cases have different expressions from D(d,p)T and D(d,n)^{3}He reactions.

Figure 14 presents ENDF/B-VIII.0 fusion cross section data obtained with accelerator-based beam-target experiments in Brookhaven National Laboratory. These fusion cross sections have been used as reference values to compare with other fusion cross-section equations (cases 1, 2 and 3).

Figures 15, 16 and 17 show fusion cross section for D+D, D+T and D+^{3}He evaluated through cases 1, 2 and 3, whereas table 6 summarizes resonance energies for all cases and fusion reactions and table 7 shows cross-section maximum values and energy E_{max}. As observed, resonance peaks are mostly attenuated in all cases and shifted from the energy at which is placed the maximum cross-section values.

Fusion cross section differences are relatively lower between all cases under study (< 5 %) in the energy range of [0.001, 5] MeV, compared with ENDF/B-VIII.0, as observed in figures 18, 19 and 20. In cases 2 and 3, differences between fusion reaction $^{3}He(d,p)^{4}He$ has several peaks at energies lower than 10 keV.

In addition, the energy position of the maximum value of fusion cross-section is also shifted between cases, as observed in table 10.

Figures 21, 22 and 23 represents comparison of fusion reactivities for D+D, D+T and D+^{3}He plasma fuels by different cross section estimations, respectively.

Figure 24 presents the power density for different ionised fuel plasmas by NRL(2019) formulation (Duane's parametrization).

Figures 25 and 26 show the ignition and break-even conditions of a fusion reactor, employing the Hively-Kozlov formulation for fusion reactivities. As observed, the minimum temperature in which fusion products overpass bremstrahlung losses is 3.6 keV for break-even and 4.5 keV for ignition. The minimum fusion triple product (FTP) is achieved for ignition at a temperature of 15.1 keV and for break-even at 14.9 keV with values $(\eta\tau T)_{ig} = 2.81\times10^{21}m^{-3}.s.keV$ and $(\eta\tau T)_{bk} = 1.37\times10^{21}m^{-3}.s.keV$, respectively (see Annex B for fusion reactor parameters)

10. Conclusions

Fusion cross sections and reactivities have been applied in nuclear technology to estimate reaction rates, probabilities and power production. During decades, fusion cross-sections have been estimated with beam-target experiments on accelerators and interpolated through several parametric equations to quantum mechanical models (R-matrix) instead of employing fusion data from fusion reactors.

In this work, several cross section and reactivity models have been compared, observing that there are considerable differences in the astrophysical S-factors formulations, derived from differences between resonance energies E_0 at the resonance peak and FWHM in all cases analysed. S-factors characterize the physics of the reaction in a more detailed way than cross sections, due to the fact that the SLBW resonance peaks at different energies are not attenuated by other factors, as occurs in the cross-sections.

In the case of tokamaks and stellerator fusion reactors, fusion cross-sections and reactivities ($<\sigma v>$) must be still reviewed due to the fact that both target and incident particles have similar kinetic and potential energies in the plasma.

In conclusion, fusion reaction models for light nuclei (deuterium, tritium and helium) must be still improved and reviewed in order to apply fusion to energy production in safety conditions.

11. References

Adelberger EG et al. Solar fusion cross sections. Rev Mod Phys 70, 1265 (1998)

Adelberger EG et al. Solar fusion cross sections II. The pp chain and CNO cycles. Rev Mod Phys 83, 195 (2011)

Arbuzov AB and Kopylova TV. On relativization of the Sommerfeld-Gamow-Sakharov factor. J High Ener Phys 9 (2012)

Atzeni S and Meyer-ter-Vehn. Nuclear fusion reactions. Oxford University Press. ISBN: 978-0-19-85624-1 (2001)

Aygun M. Comparative analysis of proximity potentials to describe scattering of ^{13}C, ^{16}O, ^{28}Si and ^{208}Pb nuclei. Rev Mex Fis E vol 64 n.2 (2018)

Ball J. Maximizing specific energy by breeding deuterium. ArXiv:1908.00834 [physics.pop-ph]
doi: 10.1088/1741-4326/ab394c (2019)

Belov AA, Kalitkin NN and Kozlitin IA. Refinement of thermonuclear reaction rates. Fusion Engineering and Design 141, 51-58 (2019)

Bethe HA. Energy production in Stars. Physical Review 55(5): 434-56 (1939)

Blatt JM and Weisskopf VF. Theoretical Nuclear Physics (Springer-Verlag, 1979)

Breit G and Wigner E. Capture of Slow Neutrons, Phys. Rev. 49, 519 (1936)

Brennan JG and Coyne JJ. Energy Dependence of the D-D reaction cross section at low Energies. J Res Natl Bur Stand A Phys Chem 68A(6): 675-682 (1964)

Bosch and Hale 1992. Improved formulas for fusion cross-sections and thermal reactivities. NUCLEAR FUSION, vol 32, N0.4 (1992)

Brown DA et al. ENDF/B-VIII.0: The 8th major release of the nuclear reaction data library with CIELO-project cross sections, new standards and thermal scattering data. Nucl Data Sheets 148(2018)1

Bystritsky VM, Grebenyuk VM, Parzhitski SS, Pen'kov FM, Sidorov VT and Stolupin VA. Experimental Investigation of dd reaction in range of ultralow energies using Z-pinch. D15-98-239. Nucl Inst Meth (1998)

Chen FF. Introduction to Plasma Physics and Controlled Fusion (vol 1) Springer (2016)
ISBN 978-3-319-22309-4

Chepurnov VA and Fiz Y. 955 (1967)

Denisov VY. Interaction potential between heavy ions. Phys Letters B 526 315-321 (2002)

Dolan TJ. Magnetic Fusion Technology. Lecture Notes in Energy. ISBN 978-1-4471-5555-3 (2013)

Donné AJH. The European roadmap towards fusion electricity. Phil. Trans. R. Soc. A 377: 20170432. (2018) doi:10.1098/rsta.2017.0432

Duane BH. Fusion Cross Section Theory, in Rept. BNWL-1685 (BattellePacific Northwest Laboratory, USA 1972)

Dudek J, Majhofer A, Skalski J, Werner T, Cwiokand S and Nazarewicz W. J. Phys.G 5,101359 (1979)

Dudek J, Szymanski Z, Werner T, Faessler A and Lima C. Phys. Rev.C261712 (1982)

Englefield MJ. Solution of Coulomb problem by Laplace Transform. J Math Anal Appl. 48, 270-275 (1974)

Ghasemi R and Sadeghi H. S-factor for radiative capture reactions for light nuclei at astrophysical energies. Results in Physics 9 151-165 (2018)

Green AES. Nuclear Physics, McGraw-Hill, New York 1955.

Greiner W. Quantum Mechanics, 2nd ed. Springer-Verlag Berlin Heidelberg NY (USA) 1993.

Heltemes TA, Moses GA and Santarius JE. Analysis of an Improved Fusion Reaction rate Model for use in Fusion Plasma Simulations. Fusion Technology Institute UWFDM-1268 (2005)

Hively LM. Convenient computational forms of Maxwellian reactivities. Nuclear Fusion 17(4):873-876 (1977)

Hively LM. A simple computational form for Maxwellian reactivities. Nuclear Technology/Fusion vol 3, 199-200 (1983)

Humblet J, Fowler WA and Zimmerman BA. Approximate penetration factors for nuclear reactions of astrophysical interest. Astron Astrophys 177, 317-325 (1987)

Iben I. Stellar Evolution within and off the main sequence. Annual Review of Astronomy and Astrophysics 5: 571-626 doi:10.1146/annurev.aa.05.090167.003035 (1967)

Jing Z, Yuan-Yong F, Shu-Hua Z, Hai-Long X, Cheng-Bo Li and Qiu-Ying M. Measurement of the astrophysical S factor for the low energy 2H(d,gamma)4He reaction. Chinese Phys C, 33, 5 (2009)

Kikuchi M, Lackner K and Quang Tran M. Fusion Physics. ISBN 978-92-0-130410-0. IAEA (2012)

Kim Y, Mack JM, Herrmann HW, Young CS et al. Determination of the deuterium-tritium branching ratio based on inertial confinement fusion implosions. Phys Rev C85, 061601(R) (2012)

Kiptily VG. On the Core Deuterium-Tritium Fuel Ratio and Temperature Measurements in DEMO. Nuclear Fusion, 55 (2) (2015) doi: 10.1088/0029-5515/55/2/023008

Knoepfel H. Tokamak Reactors for Breakeven.A Critical Study of the Near-Term Fusion Reactor Program. Pergamon (1978)

Kozlov BN. Thermonuclear reaction rates. At Ehnerg 12, 238 (1962)

Landau LD and Lifshitz EM. Quantum Mechanics (Pergamon, Oxford 1987) p572

Langenbrunner JL, Weller HR and Tilley DR. Two-deuteron radiative capture: Polarization observables at Ed ≤ 15 MeV. Phys Rev C 42, 1214 (1990)

Langenbrunner JR and Booker JM. Analytic, empirical and delta method temperature derivatives of D-D and D-T fusion reactivity formulations, as a means of verification. LA-UR-17-26143 (2017)

Lawson JD. Some criteria for a Power Producing Thermonuclear Reactor. Proc Phys Society B 70(1): 6-10 (1955) doi:10.1088/0370-1301/70/1/303

Lee M and Jung Y. Quantum shielding effects on the Gamow penetration factor for nuclear fusion reaction in quantum plasmas. Phys Plasmas 24, 014502 (2017)

Li XZ. Nuclear physics for nuclear fusion. Fusion Science and Technology (2002) doi:10.13182/FST02-A201

Li XZ et al. Fusion cross sections for fusion energy. Fusion Eng and Design 81, 1517-1520 (2006)

Li XZ, Wei QM and Liu B. A new simple formula for fusion cross-sections of light nuclei. Nucl Fusion 48, 125003 (2008)

MacFarlane JJ, Moses GA and Peterson RR. BUCKY - A 1D Radiation Hydrodynamics Code for Simulating Inertial Confinement - Fusion High Energy Density Plasmas UWFDM984 (2002)

Mott NF and Massey HS. The theory of atomic collisions. Oxford University Press, London (1965)

Miley GH, Towner H and Ivich N. Fusion cross sections and reactivities. Rept COO-2218-17 (University of Illinois, Urbana, IL, USA 1974) Np (1974) doi:10.2172/4014032

Nishitani T, Shibata Y, Tobita K and Kusama Y. Fusion gamma-ray diagnostics for D-3He experiments in JT-60U. Review Scientific Instruments 72, 877 (2001) doi.org/10.1063/1.1321004

Nocente M, Källne J, Salewski M, Tardocchi M and Gorini G. Gamma-ray emission spectrum from thermonuclear fusion reactions without intrinsic broadening. Nuclear Fusion 55, 12 (2015). doi:10.1088/0029-5515/55/12/123009

NRL Plasma Formulary (2019) https://www.nrl.navy.mil/ppd/content/nrl-plasma-formulary

Ongena J. Nuclear fusion and its large potential for the future world energy supply. Nukleonika; 61(4): 425-432. (2016) doi:10.1515/nuka-2016-0070

Otuka et al. Towards a More Complete and Accurate Experimental Nuclear Reaction Data Library (EXFOR): International Collaboration Between Nuclear Reaction Data Centres (NRDC) Nucl Data Sheets 120, 272-276 (2014)

Peres AJ. J Nucl Mater 50, 5569 (1979)

Rajbongshi T and Kalita K. Systematic study of deformation effects on fusion cross-sections using various proximity potentials. Open Physics 12 (6) (2014) doi:10.2478/s11534-014-0451-1

Ramos-Pascual M. Fusion cross-sections for deuterium cycle fusion reactors (D-cycle): an analysis of geometric, Gamow-Sommerfeld and astrophysical S-factors. OSF preprints. doi:10.31219/osf.io/aur4e

Ran Y, Xue L, Hu S and Su RK. On the Coulomb-type potential of the one-dimensional Schrödinger equation. J Phys A: Math Gen 33 9265-9272 (2000)

Robouch BV et al. Gamma diagnostics of thermonuclear plasma using D(d,gamma)^{4}He reaction: Feasibility study. ETDE-IT-94-23 IAEA (1993)

Sadler G and van Belle P. An improved formulation of the D(t,n)^{4}He reaction cross-section. Tech Rep JET-IR(87)08. JET Joint Undertaking (1987)

Singh V, Atta D, Khan MA and Basu DN. Astrophysical S-factor for deep sub-barrier fusion reactions of light nuclei. Nuclear Physics A (986) 98-106 (2019)

Smith CL and Cowley S. The path to fusion power. Phyl Trans R Soc 368, 1091-1108 (2010) doi:10.1098/rsta.2009.0216

Takano H and Ishiguro Y. Multi-level correction to Breit-Wigner Single-Level Formula. Jour Nucl Sci and Tech 14(9), 627-639 (1977)

Woods LC. Theory of Tokamak Transport (Wiley-VCH, 2006)

Wurzel SE and Hsu SC. Progress toward fusion energy breakeven and gain as measured against the Lawson criterion. Arxiv (2021) //arxiv.org/abs/2105.10954

Yoon JH and Wong CY. Relativistic modification of the Gamow factor. Phys Rev C 61, 044905 (2000) doi:10.1103/PhysRevC.61.044905

Zhom H. On the size of tokamak fusion power plants. Phil. Trans. R. Soc. A 377: 20170437. doi:10.1098/rsta.2017.0437

Zhong-Li X, Liu B, Chen S, Ming Wei Q and Hora H. Fusion cross-sections for inertial fusion energy. Laser and Particle Beams 22, 469-477 (2004) doi:10.1017/S026303460404011X

Tables

Table 1 - Fusion reactions of the D-cycle between deuterium and other light nuclei [Langenbrunner et al 1990, Jing et al 2009, Kiptily 2014, Ongena 2016]

	Reaction		BR_i	E_f (MeV)	Energy released per reaction product [MeV]		
					(a)	(b)	(c)
The D-cycle	D+p	^{3}He+γ	1	5.49	-	5.49	-
	D+D	T+p	0.45	4.03	1.01	3.02	-
		^{3}He+n	0.55	3.27	0.82	2.45	-
		^{4}He+γ	10^{-7}-10^{-6}	23.85	0.08	23.77	-
	D+T	^{3}He+2n	-	-2.98	-	-	-
		^{4}He+n	~1	17.59	3.52	14.07	-
		^{5}He+γ$^{(a)}$	$5x10^{-5}$ - $5x10^{-4}$	16.70	0.05 / 3.2	16.65 / 13.50	-
	D+^{3}He	^{3}He+p+n	-	-2.22	-	-	-
		^{4}He+p	~1	18.35	3.67	14.68	-
		^{5}Li+γ$^{(b)}$	$5x10^{-5}$ - $5x10^{-4}$	16.39	0.05	16.34	-
The T+T reaction	T+T	^{4}He+2n	1	11.33	3.79	3.76	-
Advanced fusion fuels	T+^{3}He	^{4}He+p+n	0.51	12.10	2.02	5.04	5.03
		^{4}He+D	0.43	14.32	4.79	9.52	-
		^{5}He+p	0.06	11.21	1.87	9.33	-
Neutron fusion	n+p	D+γ	1	2.22	-	2.22	-

(a) D+T branching ratio by inertial confinement fusion (ICF) [Kim et al 2012]

(b) D+^{3}He branching ratio on JT-60U tokamak [Nishitani et al 2000]

Table 2 - Fusion cross-section parameters for Duane's 5 parameter equation [Miley et al 1974, NRL 2019]

	D(d,p)T	D(d,n)^{3}He	T(d,n)^{4}He	^{3}He(d,p)^{4}He	Units	Comments
A1	46.097	47.88	45.95	89.27	keV$^{1/2}$	Gamow constant (B_G)
A2	372	482	5.02 x 10^4	2.59 x 10^4	keV.b	-
A3	4.36 x 10^{-4}	3.08 x 10^{-4}	1.368 x 10^{-2}	3.98 x 10^{-3}	keV	-
A4	1.22	1.177	1.076	1.297	-	-
A5	0	0	409	647	keV.b	-
E_0	2798.16	3821.43	78.65	325.88	keV	Resonance energy E_0=A4dd1/A3dd1
Γ_a	4587.16	6493.51	146.20	502.51	keV	Resonance width Γ_a=2/A3dd1

Table 3 - S-function (Padé expansion coefficients) [Bosch and Hale 1992]

	D(d,p)T	D(d,n)^{3}He	T(d,n)^{4}He		^{3}He(d,p)^{4}He		Units
			[0.5-550 keV]	[550-4700 keV]	[0.3-900 keV]	[900-4800 keV]	
B_g	31.3970	31.3970	34.3827	34.3827	68.7508	68.7508	keV$^{1/2}$
A1	5.5576 x 10^4	5.3701 x 10^4	6.927 x 10^4	-1.4714 x 10^6	5.7501 x 10^6	-8.3993 x 10^5	keV.mb
A2	2.1054 x 10^2	3.3027 x 10^2	7.454 x 10^8	0.0	2.5226x 10^3	0.0	mb
A3	-3.2638 x 10^{-2}	-1.2706 x 10^{-1}	2.050 x 10^6	0.0	4.5566x 10^1	0.0	mb.keV^{-1}
A4	1.4987 x 10^{-6}	2.9327 x 10^{-5}	5.2002 x 10^4	0.0	0.0	0.0	mb.keV^{-2}
A5	1.8181 x 10^{-10}	-2.5151 x 10^{-9}	0.0	0.0	0.0	0.0	mb.keV^{-3}
B1	0.0	0.0	6.38 x 10^1	-8.4127 x 10^{-3}	-3.1995 x 10^{-3}	-2.6830 x 10^{-3}	keV^{-1}
B2	0.0	0.0	-9.95 x 10^{-1}	4.7983 x 10^{-6}	-8.5530 x 10^{-6}	1.1633 x 10^{-6}	keV^{-2}
B3	0.0	0.0	6.981 x 10^{-5}	-1.0748 x 10^{-9}	5.9014x 10^{-8}	-2.1332 x 10^{-10}	keV^{-3}
B4	0.0	0.0	1.728 x 10^{-4}	8.5184 x 10^{-14}	0.0	1.4250 x 10^{-14}	keV^{-4}

Table 7 - Averaged nuclear radius r_{12}, Wood-Saxon nuclear potential well $U_S = U_r + iU_i$, Coulomb barrier U_C and factor β [Li et al 2008, Singh et al 2018]

	D(d,p)T / D(d,n)^{3}He	T(d,n)^{4}He	^{3}He(d,p)^{4}He	Units
r_0	2.778	1.887	3.331	fm
r_{12}	7.0	5.1	9.0	fm
U_r	-48.52	-40.69	-11.859	MeV
U_i	-263.27	-109.18	-259.02	keV
U_C	205.71	282.42	319.97	keV
μ	1.6737 x 10^{-27}	2.0087 x 10^{-27}	2.0083 x 10^{-27}	kg
β	$2\sqrt{2}\pi$	$\sqrt{\pi}$	$2\sqrt{2}$	-

Table 5 - Comparison of geometric, Gamow and S-factors of several cases of fusion cross-section expressions

Case	Φ(E)	G(E)	S(E)	Comments	Reference
1	$1/E$	$\left(\exp\left[\frac{A1}{\sqrt{E}}\right]-1\right)^{-1}$	$A_5+\left(\frac{A_2}{(A_4-A_3E)^2+1}\right)$	Duane's 5 parameter formula with a [2/2] order Padé approximation of S(E)	[NRL 2019]
2	$1/E$	$\left(\exp\left[\frac{B_G}{\sqrt{E}}\right]\right)^{-1}$	$\frac{A_1+E(A_2+E(A_3+E(A_4+EA_5)))}{1+E(B_1+E(B_2+E(B_3+EB_4)))}$	Bosch-Hale formula with a [4/4] order Padé approximation of S(E)	[Bosch and Hale 1992]
3	π/k^2	$2\pi\left(\exp\left[\frac{2\pi}{ka_c}\right]-1\right)^{-1}$	$\left(\frac{-4\omega_i}{\omega_r^2+[\omega_i-(1/\chi^2)]^2}\right)$	Woods-Saxon square complex potential with phase shift $\omega(\delta_0)$ boundary conditions	[Li et al 2006, 2008]

Table 6 - Fusion reactivity <σv> in cm^3s^{-1} [NRL]

T(keV)	D(d,p)T + D(d,n)^{3}He	T(d,n)^{4}He	^{3}He(d,p)^{4}He
1.0	1.5×10^{-22}	5.5×10^{-21}	10^{-26}
2.0	5.4×10^{-21}	2.6×10^{-19}	1.4×10^{-23}
5.0	1.8×10^{-19}	1.3×10^{-17}	6.7×10^{-21}
10.0	1.2×10^{-18}	1.1×10^{-16}	2.3×10^{-19}
20.0	5.2×10^{-18}	4.2×10^{-16}	3.8×10^{-18}
50.0	2.1×10^{-17}	8.7×10^{-16}	5.4×10^{-17}
100.0	4.5×10^{-17}	8.5×10^{-16}	1.6×10^{-16}
200.0	8.8×10^{-17}	6.3×10^{-16}	2.4×10^{-16}
500.0	1.8×10^{-16}	3.7×10^{-16}	2.3×10^{-16}
1000.0	2.2×10^{-16}	2.7×10^{-16}	1.8×10^{-16}

Table 7 - Fusion reactivity <σv> coefficients [Bosch and Hale 1992]

	D(d,p)T	D(d,n)^{3}He	T(d,n)^{4}He	^{3}He(d,p)^{4}He	Units
B_g	31.3970	31.3970	34.3827	68.7508	keV$^{1/2}$
μc^2	937814	937814	1124656	1124572	keV
C1	5.65718×10^{-12}	5.43360×10^{-12}	1.17302×10^{-9}	5.51036×10^{-10}	
C2	3.41267×10^{-3}	5.85778×10^{-3}	1.51361×10^{-2}	6.41918×10^{-3}	keV^{-1}
C3	1.99167×10^{-3}	7.68222×10^{-3}	7.51886×10^{-2}	-2.02896×10^{-3}	keV^{-1}
C4	0.0	0.0	4.60643×10^{-3}	-1.91080×10^{-5}	keV^{-2}
C5	1.05060×10^{-5}	-2.96400×10^{-6}	1.35000×10^{-2}	1.35776×10^{-4}	keV^{-2}
C6	0.0	0.0	-1.06750×10^{-4}	0.0	keV^{-3}
C7	0.0	0.0	1.36600×10^{-5}	0.0	keV^{-3}

Table 8 - Fusion reactivity <σv> coefficients of Hively-Kozlov formulation [Hively 1983]

	D(d,p)T	D(d,n)^{3}He	T(d,n)^{4}He	^{3}He(d,p)^{4}He
A1	$1.8742387 \times 10^{-14}$	$2.6001260 \times 10^{-14}$	$3.3729290 \times 10^{-12}$	$-4.2145988 \times 10^{-14}$
A2	19.244282	19.751844	20.407142	28.415670
A3	1.6109164×10^{-2}	1.2703127×10^{-2}	0.20293722	-2.5739067
A4	-	-	8.8829520×10^{-5}	1.3007516×10^{-6}

Table 9 - Comparison of resonance parameters for the D-cycle fusion reactions

		Case 1	Case 2	Case 3	Units
E_0	D(d,p)T	2798.2	-	931.3	keV
	D(d,n)^{3}He	3821.4	5012.7		
	T(d,n)^{4}He	78.7	49.10	42	
	^{3}He(d,p)^{4}He	325.9	215.3	207.8	
Γ_a *(FWHM)*	D(d,p)T	4587.2	-	846.4	keV
	D(d,n)^{3}He	6493.5	5517.9	846.4	
	T(d,n)^{4}He	147.4	81.2	82.1	
	^{3}He(d,p)^{4}He	515.2	287.55	172.7	
S(0) at 1 *keV*	D(d,p)T	0.14	0.05	0.04	MeV.b
	D(d,n)^{3}He	0.20	0.05	0.04	
	T(d,n)^{4}He	23.99	11.71	4.25	
	^{3}He(d,p)^{4}He	10.34	5.77	1.23	
S(E) at E_0	D(d,p)T	0.37	-	0.21	MeV.b
	D(d,n)^{3}He	0.48	0.62	0.21	
	T(d,n)^{4}He	50.61	27.29	8.47	
	^{3}He(d,p)^{4}He	26.55	16.69	8.19	

Table 10 - Comparison of fusion cross section parameters for the D-cycle fusion reactions

		ENDF	Case 1	Case 2	Case 3	Units
E_{max}	D(d,p)T	1700	2550	7999	912	keV
	D(d,n)^{3}He	1200	3400	1109	912	
	T(d,n)^{4}He	64.8	113.3	64.8	62.3	
	^{3}He(d,p)^{4}He	247.8	430.4	262.3	230.1	
σ_f *at* E_{max}	D(d,p)T	8.9×10^{-2}	9.6×10^{-2}	1.0×10^{-1}	8.9×10^{-2}	barn
	D(d,n)^{3}He	0.104	0.109	0.105	0.089	
	T(d,n)^{4}He	5.01	4.94	5.04	5.00	
	^{3}He(d,p)^{4}He	0.81	0.72	0.81	0.80	

Figures

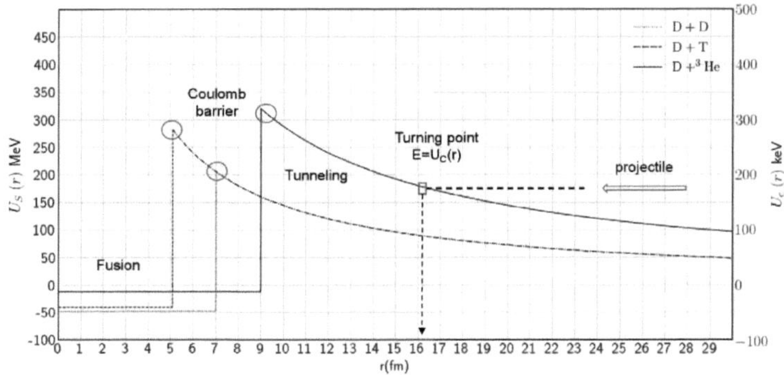

Figure 1 - Woods-Saxon square-well (left) and Coulomb electrostatic proximity potential (right) for fusion reactions

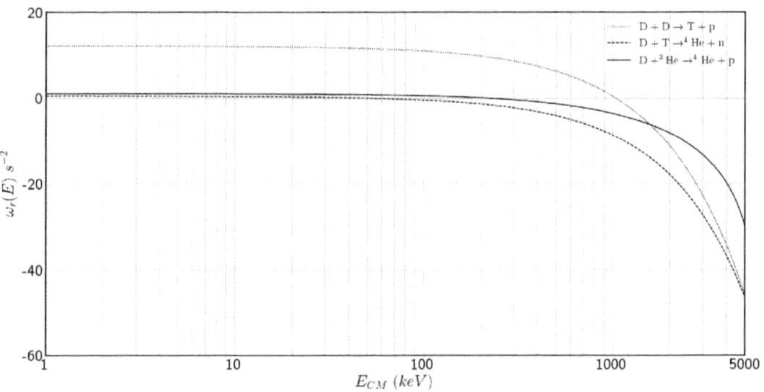

Figure 2 - Complex phase shift $\omega_r(E)$ of fusion reactions (Re)

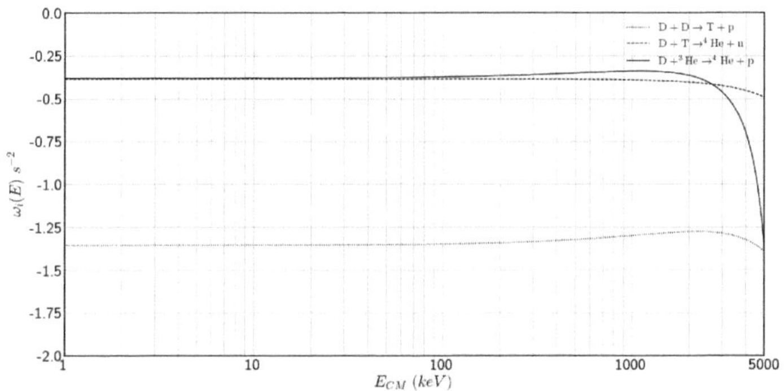

Figure 3 - Complex phase shift $\omega_i(E)$ of fusion reactions (Im)

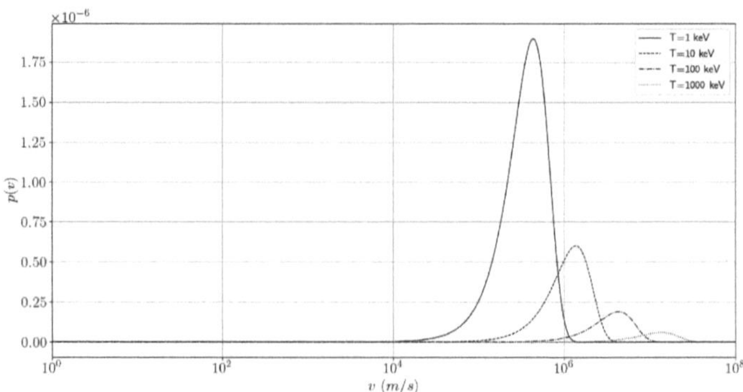

Figure 4 - Distribution function *p(v)* of velocities in a D+D ionised plasma at different temperatures (Maxwell-Boltzmann distribution)

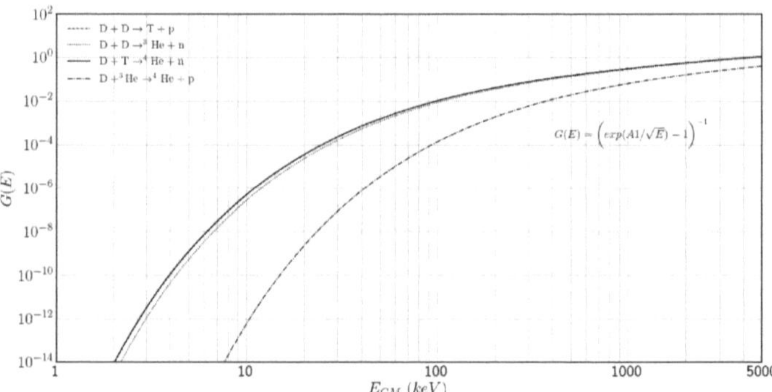

Figure 5 - Gamow factors as a function of center-of-mass energy *E* from Duane's 5 parameter equation [NRL 2019]

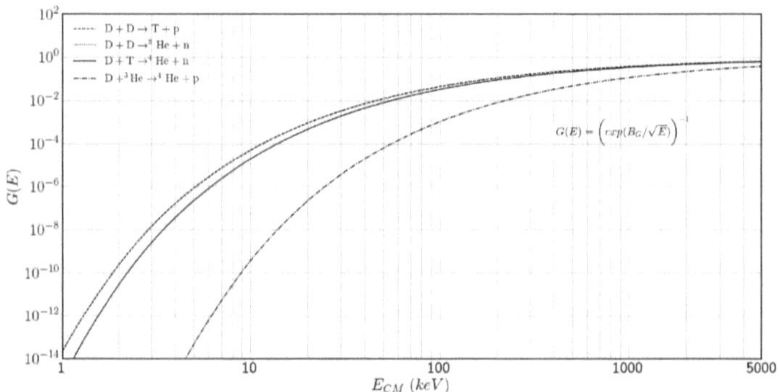

Figure 6 - Gamow factors as a function of center-of-mass energy *E* (reduced Mott's expression)

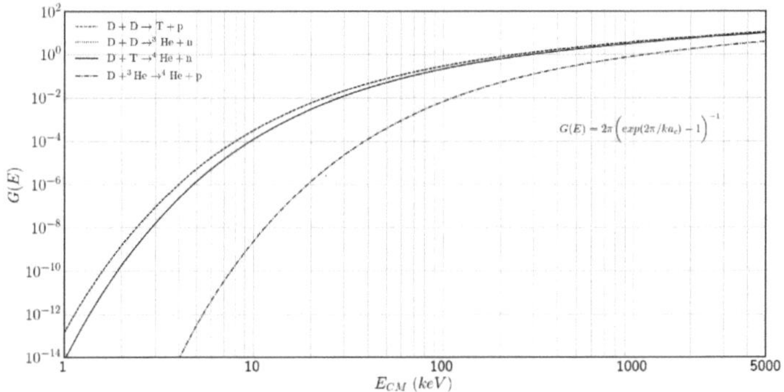

Figure 7 - Gamow factors ($1/\chi^2$) as a function of center-of-mass energy *E*

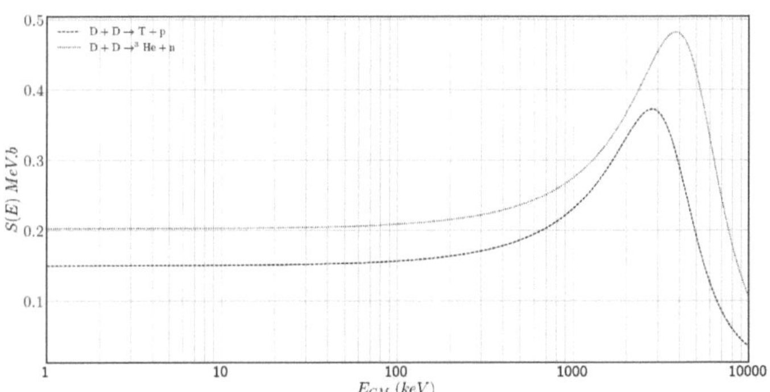

Figure 8 - S-factor of D+D fusion reactions as a function of center-of-mass energy *E* (Case 1)

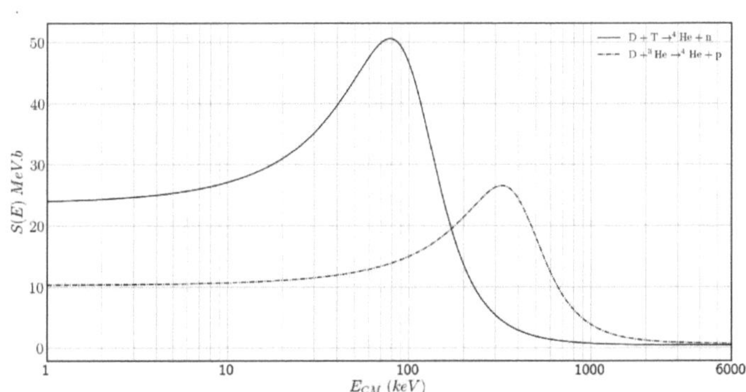

Figure 9 - S-factor of D+T and D+^{3}He fusion reactions as a function of center-of-mass energy *E* (Case 1)

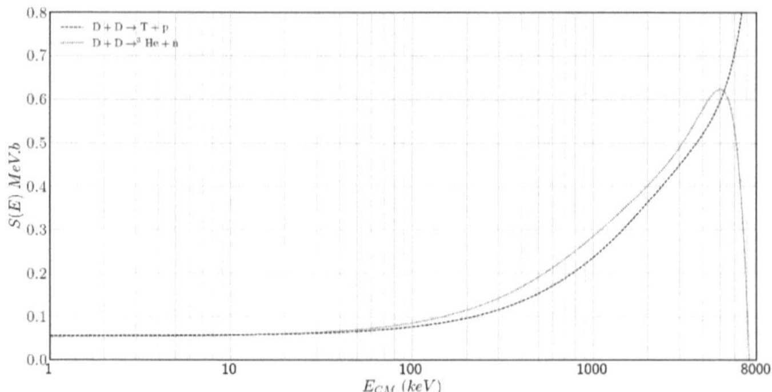

Figure 10 - S-factor for D+D fusion reaction as a function of center-of-mass energy *E* (Case 2)

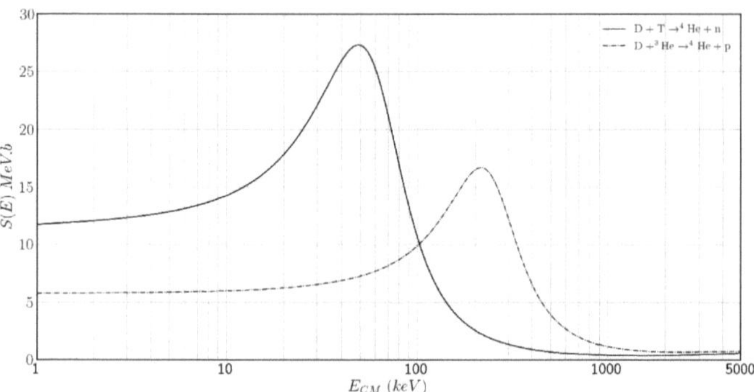

Figure 11 - S-factor for D+T and D+^{3}He fusion reactions as a function of center-of-mass energy *E* (Case 2)

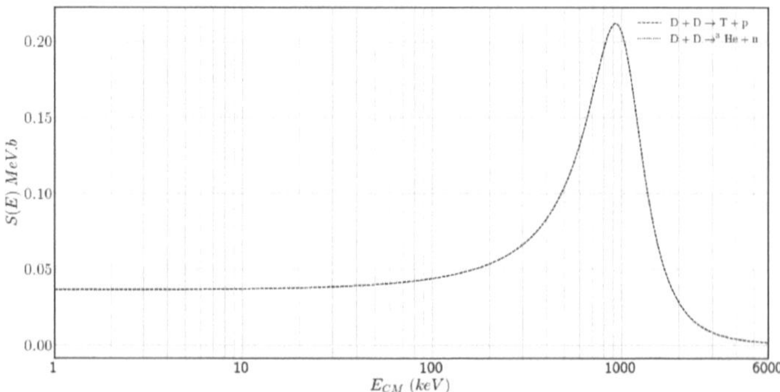

Figure 12 - S-function for D+D fusion reaction as a function of center-of-mass energy *E* (Case 3)

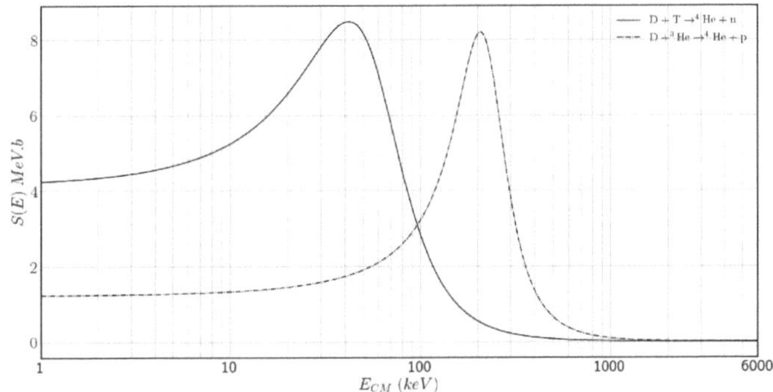

Figure 13 - S-function for D+T/D+^{3}He fusion reactions as a function of center-of-mass energy *E* (Case 3)

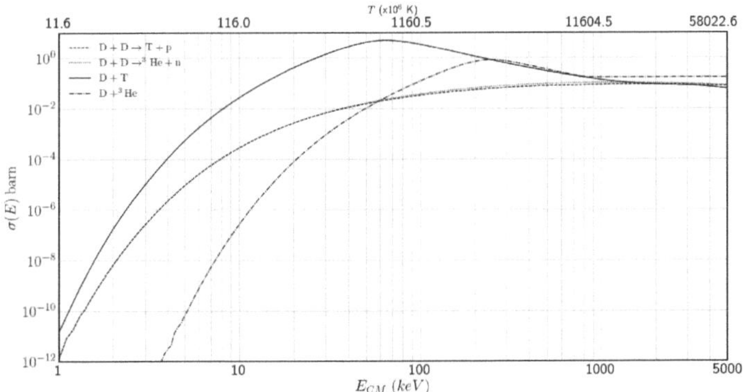

Figure 14 - Fusion cross-sections for D+D/D+T/D+^{3}He reactions as a function of center-of-mass energy *E* from ENDF/B-VIII.0 (Reference)

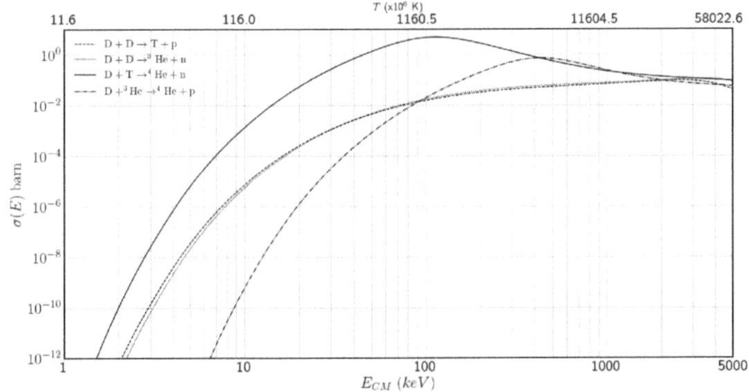

Figure 15 - Fusion cross-sections for D+D/D+T/D+^{3}He reactions as a function of center-of-mass energy *E* (Case 1)

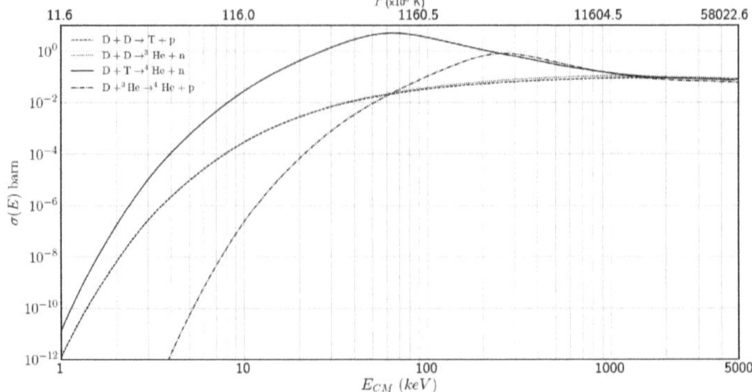

Figure 16 - Fusion cross-sections for D+D/D+T/D+^{3}He reactions as a function of center-of-mass energy *E* (Case 2)

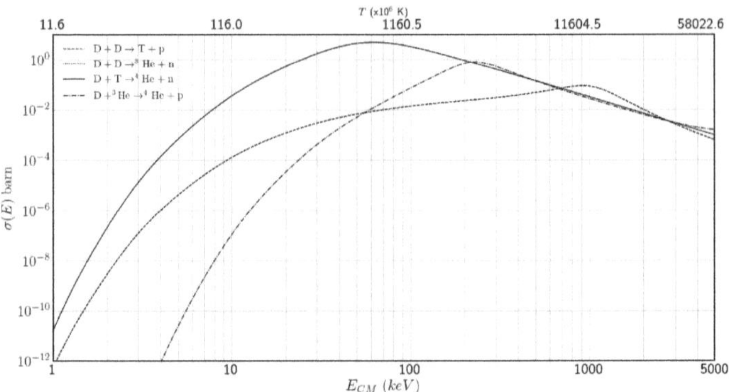

Figure 17 - Fusion cross-sections for D+D/D+T/D+^{3}He reactions as a function of center-of-mass energy *E* (Case 3)

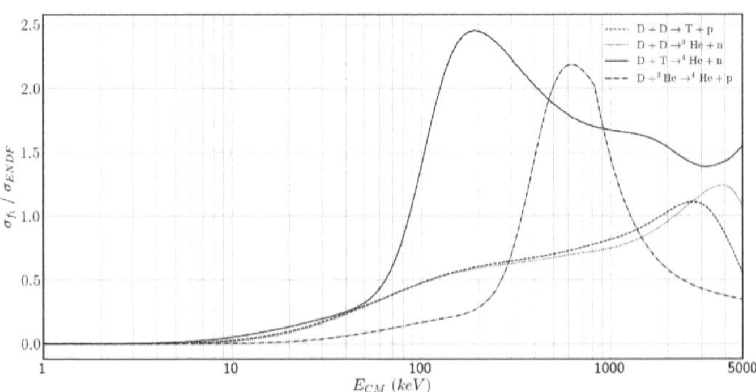

Figure 18 - Relative differences of fusion cross-sections between case 1 formula and ENDF

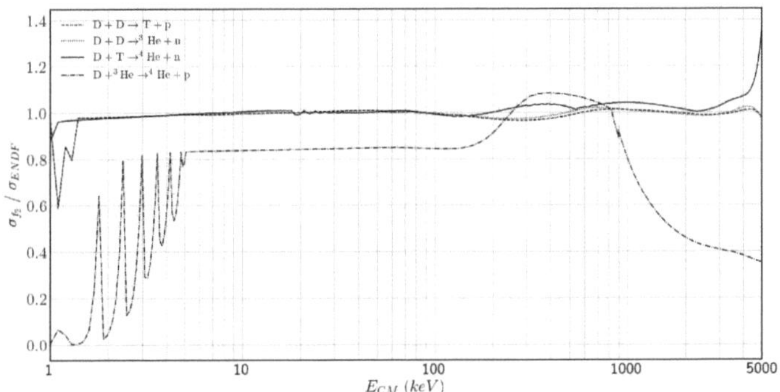

Figure 19 - Relative differences of fusion cross-sections between case 2 and ENDF

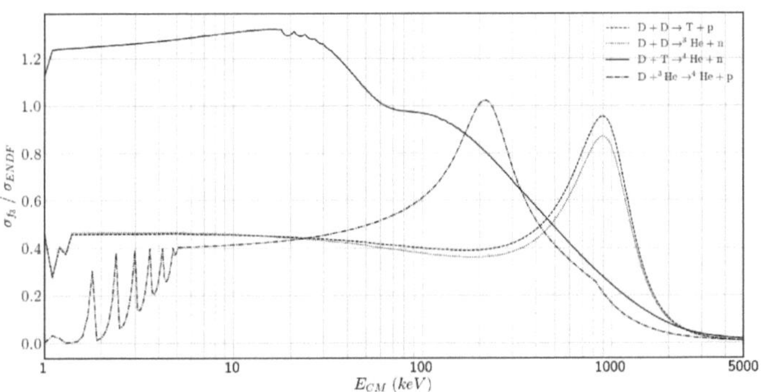

Figure 20 - Relative differences of fusion cross-sections between case 3 formula and ENDF

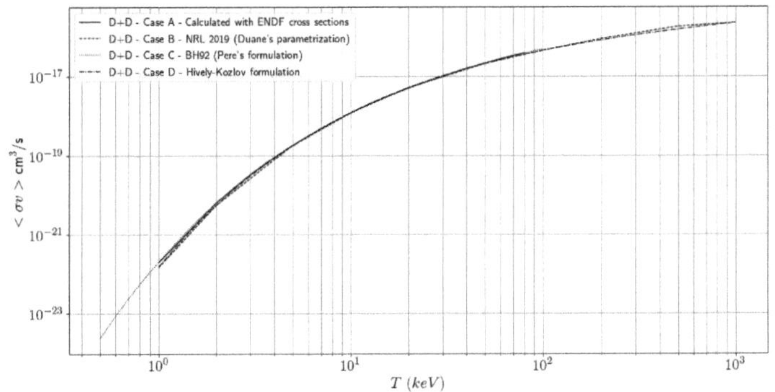

Figure 21 - Comparison of fusion reactivities <σv> in a D+D ionised plasma

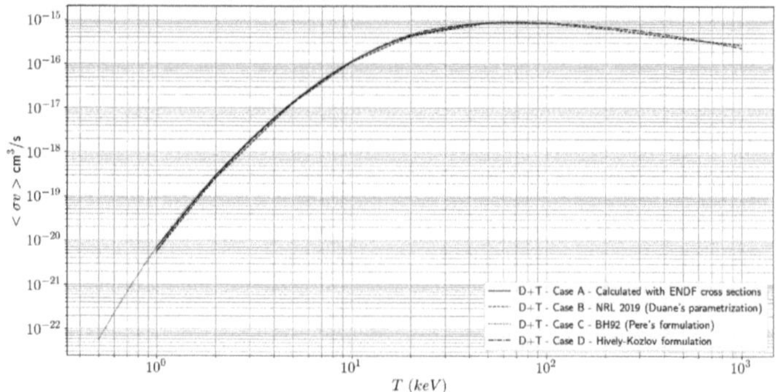

Figure 22 - Comparison of fusion reactivities <σv> in a D+T ionised plasma

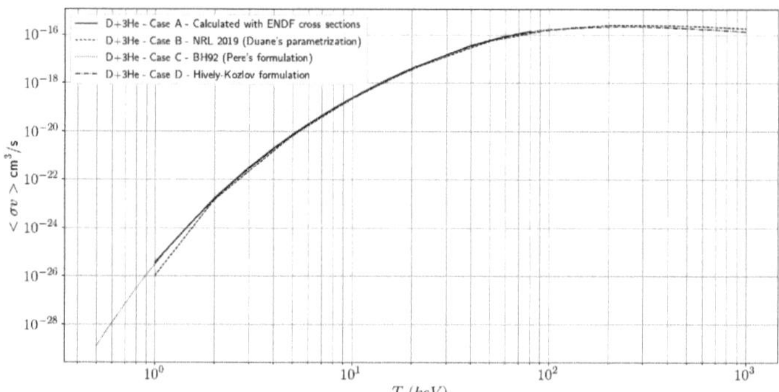

Figure 23 - Comparison of fusion reactivities <σv> in a D+^{3}He ionised plasma

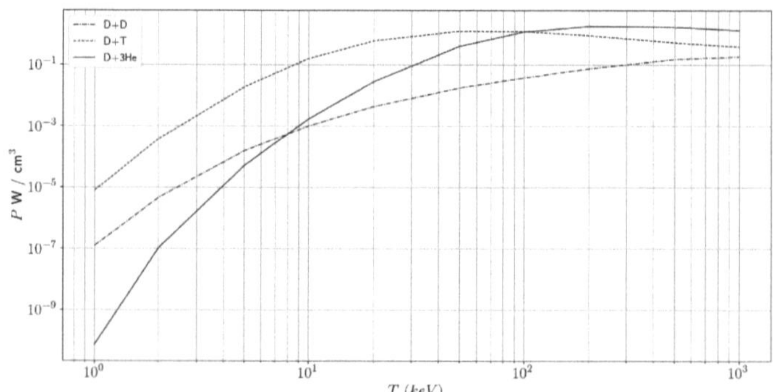

Figure 24 - Power densities in different ionised plasmas by NRL(2019) - Duane's parametrization

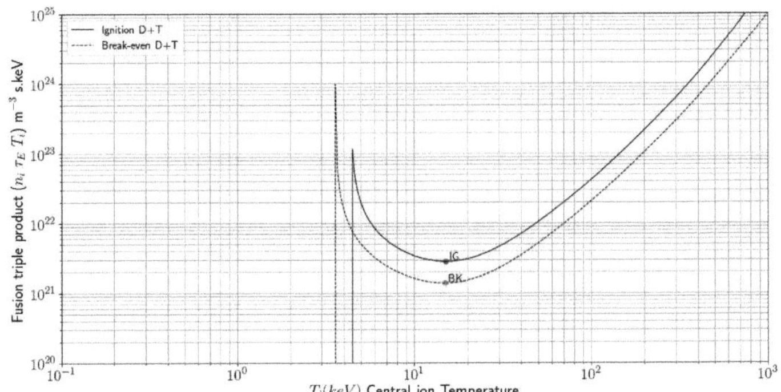

Figure 25 - Fusion triple product (FTP) of break-even and ignition conditions for a D+T plasma fusion reactor

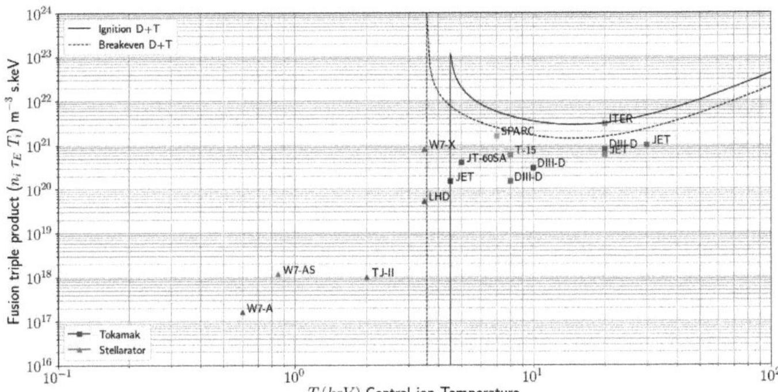

Figure 26 - Comparison of FTP under nominal operation conditions of several fusion reactors (tokamaks and stellarators)

Annex A: The SIMURGH code

The SIMURGH code is a software to simulate fusion reaction conditions of plasma in D-cycle and tritium in tokamaks and stellerators. In particular, it estimates fusion cross sections, fusion reactivities, geometric, Gamow-Sommerfeld, S-factors, energy resonances and partial widths (FWHM) in the deuterium cycle (D-cycle) and energy range from 1 keV to 5 MeV, equivalent to plasma temperatures of 11.6 x 10^6 to 5.80 x 10^{10} K.

The software has been developed in Python 3.8 environment and includes ENDF fusion cross section library. The SIMURGH code calculates also the reaction heat (Q), the energy of each product and the parameters of the Woods-Saxon and Coulomb proximity potentials of each type of fusion reactions, that will be subject of further research (see figures A1 and A2).

Fusion reactivities have been estimated as

$$< \sigma \upsilon >= \frac{\left[\sum_{k=0}^{N-1} v_k p(v_k) \sigma(E_k) + v_{k+1} p(v_{k+1}) \sigma(E_{k+1}) \right]}{\left[\sum_{k=0}^{N-1} p(v_k) + p(v_{k+1}) \right]}$$

The number of points for the integration has been fixed in N=10^3 points, and the minimum and maximum limits for the Maxwell-Boltzmann distribution of speeds in $v_{min}=0$ *m/s*, $v_{max}=10^8$ m/s, with a step interval of $\Delta = (v_{\max} - v_{\min})/N$ = 10^3 m/s.

```
##########################################################################################
#
# SIMURGH - SIMU-lations in fusion technology
#           A fusion reactor software for plasma reactions in tokamaks and stellerators
#
##########################################################################################

Simulation time:  2021-10-30 19:41:04.301393

 |-------------------------------------------------------------------------|
 |                                                                         |
 |     [0] Physical constants:                                             |
 |                                                                         |
 |-------------------------------------------------------------------------|

Physical constants:
 mp= 1.67353269131030 8e-27  kg (proton mass)
 mn= 1.6749273500334555e-27  kg (neutron mass)
  e= 1.602176634e-19  C (electron charge (C) = 1 eV in J)
 kb= 1.380649e-23  J.K-1 (Boltzmann constant)
 kg= 8.314462618  J.mol-1.K-1 (Ideal gas constant)
 alpha= 0.0072973525692 7803  (fine structure constant)
 hbar= 1.0545718176461565e-34 kg m2.s-1 (reduced Planck constant)
 c= 299792458.0 m.s-1 (speed of light)


Reduced mass:

 mu(D+D)= 1.6742300210718818e-27 kg
 mu(D+T)= 2.0091875887547003e-27 kg
 mu(D+3He)= 2.0089644432294482e-27 kg

 |-------------------------------------------------------------------------|
 |                                                                         |
 |     [1] Reaction heat (Q):                                              |
 |                                                                         |
 |-------------------------------------------------------------------------|
```

Figure A1 -SIMURGH code for fusion technology

The SIMURGH code has been applied to simulate a simplified fusion reactor of $L=10^{-8}$ m and $n_i=50\text{x}10^{13}$ ions/cm^3 (deuterons, tritions and 3He ions in equimolar concentrations) by Monte Carlo techniques. Simulation has been simplified considering no bremstrahlung radiation losses and no scattering between particles. Each particle that fuses is removed from the simulation, and also the fusion products, that escape from the cell. Fusion reactions have been estimated through ENDF cross sections (fig. A3). Table A1 shows the particle concentration in a tokamak after N=1 cycle of the Monte Carlo simulation.

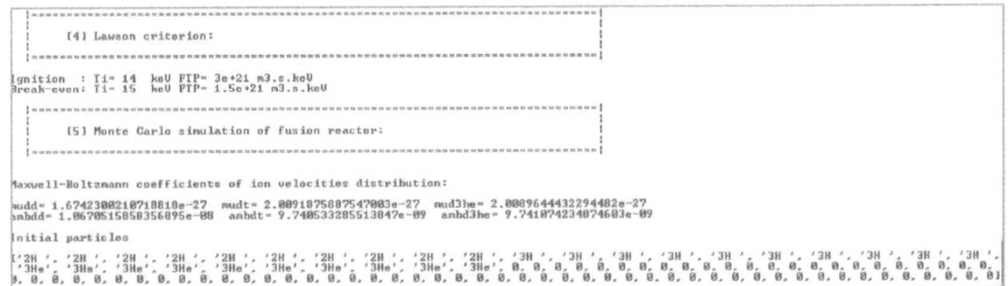

Figure A2 -Example of Lawson criterion estimation and Monte Carlo simulation

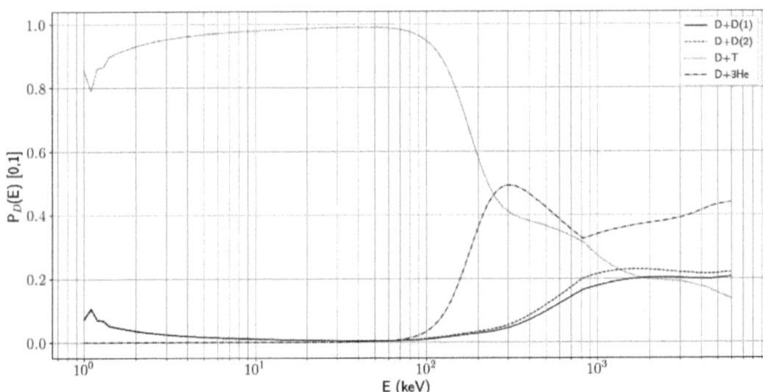

Figure A3 -Fusion probabilities P(E) for different fuel plasmas with ENDF-VII cross sections

Table A1 - Particle concentrations in tokamak after N=1 cycle

	t_0	t_1
^2D	1/3	0
^3T	1/3	0.04
^3He	1/3	0.06
^4He	-	0.4
^0n	-	0.19
^1H	-	0.31

The SIMURGH code has been tested to be a reliable and efficient software to simulate different models of fusion reactions and estimate cross sections, reactivities (<σv>) and reaction heat (Q). In the future, the code will be enlarged to include Magneto-hydrodynamic (MHD) simulations of plasma fuels and Monte Carlo simulations.

Annex B: Fusion reactor characteristic parameters

Fusion reactor type	Stellarator			Tokamak					
	Spain	Japan	Germany	USA		Japan	Russia	UK	France
Fusion reactor model	TJ-II	LHD	W 7-X	DIII-D	SPARC	JT-60SA	T15[(1)]	JET	ITER
Plasma fuel	D	D	D,T	DD	D+T	D+T	D+T	DD/DT	D+T
Plasma divertor [(2)]	LSI	HTH	LSI	SN, DN		SN, DN	SN	SN, DN	SN, DN
Major radius (R_0)	1.5	3.5-3.9	5.5	1.67	1.85	3.06	2.43	2.96	6.2
Minor radius *a* (horizontal) [m]	0.2	0.6	0.53	0.67	0.57	0.96/1.18	0.7	3.0	2.0
Minor radius *a* (vertical) [m]	-	-	-	1.74	-	2.30	-	2.1	3.7
Toroidal magnetic field B_t [T] - B_{max} [T]	1	2-3	3	2.2	12.2	2.68	3.5	3.5	5.3 - 12-19
Plasma current I_p ($x10^6$ A)	0.001	-	-	[1-3]	8.7	5.5	1.4	5.0	15(17)
Pulse length [s]	< 0.2	-	1800	10	10	100	5	60	300-500
Plasma Temperature [keV]	2	10	3.5	[8-12]	7	[2-12]	[5-10]	[8-50]	20
Plasma density [x 10^{13} cm^{-3}] - line averaged / on-axis	1	1.5	8	2.3 / 3.4	30	4.0	[3-14]	6.0	10
Energy Confinement time τ_E(s)	0.005	0.1	0.2	0.3	0.77	0.75	0.3	1.2	3
FTP ($nT\tau$) x10^{20} keV.s/m^3	0.01	0.5	0.52	[0.9-7]	16	[1-10]	[1-15]	[2-9]	30
Stored plasma energy W_p (MJ)	0.001	1.6	0.5	4	100	9.5	-	14	320
ECRH [MW]	0.6	2	10	[3-9]	-	7	1.5	38	-
ICRH [MW]	-	3	3	4	-	-	-	-	-
NBI [MW]	2	15	5	20	30	34	0.6	30	73(130) in total
Heating power [MW]	2	15	14	23	25	3	20	24	50

[(1)] Although the last prototype is the T15MD hybrid fusion/fission reactor, T15 parameters have been used for comparison purposes

[(2)] Divertor configuration: SN (Single-null), DN (Double-null), LSI (Low-Shear Stellerator Island), HTH (Helitron/Torsatron Helical)

[(3)] FTP (Fusion Triple Product, $nT\tau$)

[(4)] ECRH (Electron Cyclotron Resonance Heating), ICRH (Ion Cyclotron Resonance Heating), NBI (Neutral-Beam Injection)

Annex C: Physical constants

		Units
Electron charge	$e = 1.6022 \times 10^{-19}$	C
Electron mass	$m_e = 9.1094 \times 10^{-31}$	kg
Proton mass	$m_p = 1.6726 \times 10^{-27}$	kg
Neutron mass	$m_n = 1.674 \times 10^{-27}$	kg
Coulomb constant	$k_e = \frac{1}{4\pi\varepsilon_0} = 8.987 \times 10^9$	kg.m^3.s^{-2}.C^{-2}
Boltzmann constant	$k_b = 1.3807 \times 10^{-23}$	m^2 kg s^{-2} K^{-1} J. K^{-1}
Ideal gas constant	$k_g = k_b N_{avg} = 8.31$	J.K^{-1}.mol^{-1}
Planck constant	$h = 6.6261 \times 10^{-34}$	kg m^{2}s^{-1}
Reduced Planck constant	$\hbar = h/2\pi$	kg m^{2}s^{-1}
Fine-structure constant	$\alpha = k_e \frac{e^2}{\hbar c}$	-
Vacuum permittivity	$\varepsilon_0 = \frac{1}{\mu_0 c^2} = 8.854 \times 10^{-12}$	F.m^{-1} N.m^2.C^{-2}
Permeability free-space	μ0=4π x1.00000000082(20) 10^{-7}	H.m^{-1} N.A^{-2}
Avogadro number	$N_{avg} = 6.022 \times 10^{23}$	mol^{-1}

Annex D: Vector operators

Gradient	∇f	$\frac{\partial f}{\partial x}\vec{i}+\frac{\partial f}{\partial y}\vec{j}+\frac{\partial f}{\partial z}\vec{k}$
Laplacian	$\Delta f=\nabla^2 f=(\nabla\cdot\nabla)f$	$\frac{\partial^2 f}{\partial x^2}+\frac{\partial^2 f}{\partial y^2}+\frac{\partial^2 f}{\partial z^2}$
Divergence	$\nabla\cdot\vec{F}$	$\frac{\partial f_x}{\partial x}+\frac{\partial f_y}{\partial y}+\frac{\partial f_z}{\partial z}$
Curl (rotational)	$\nabla\times\vec{F}$	$\left(\frac{\partial f_z}{\partial y}-\frac{\partial f_y}{\partial z}\right)\vec{i}+\left(\frac{\partial f_x}{\partial z}-\frac{\partial f_z}{\partial x}\right)\vec{j}+\left(\frac{\partial f_y}{\partial x}-\frac{\partial f_x}{\partial y}\right)\vec{k}$.

- Vector product: $\vec{u}\cdot\vec{v}=\sum_i u_i v_i=\|\vec{u}\|\|\vec{v}\|\cos\theta$
- Curl product: $\vec{v}\times(\nabla\times\vec{a})=\nabla_a(\vec{v}\cdot\vec{a})-\vec{v}\cdot\nabla\vec{a}$

- $\vec{v}\cdot\nabla\vec{v}=\nabla\left(\frac{\|\vec{v}\|^2}{2}\right)+(\nabla\times\vec{v})\times\vec{v}$

- $\nabla(fg)=f\nabla g+g\nabla f$
- $\nabla(\vec{u}\cdot\vec{v})=(\vec{u}\cdot\nabla)\vec{v}+(\vec{v}\cdot\nabla)\vec{u}+\vec{u}\times(\nabla\times\vec{v})+\vec{v}\times(\nabla\times\vec{u})$
- $\nabla\cdot(f\vec{v})=(\nabla f)\cdot\vec{v}+f(\nabla\cdot\vec{v})$
- $\nabla\cdot(\vec{u}\times\vec{v})=\vec{v}\cdot(\nabla\times\vec{u})-\vec{u}\cdot(\nabla\times\vec{v})$
- $\nabla\times(\vec{u}\times\vec{v})=\vec{u}(\nabla\cdot\vec{v})-\vec{v}(\nabla\cdot\vec{u})+(\vec{v}\cdot\nabla)\vec{u}-(\vec{u}\cdot\nabla)\vec{v}$

Annex E: Magnetohydrodynamic (MHD) equations (time-dependent)

Definitions

Φ	Electric potential	V
A	Magnetic potential	T/m
ψ	Gravitational potential (gh)	J/kg
η_m	magnetic diffusivity ($1/\sigma_0 \mu_0$)	m^2/s
q	charge	C
ρ_e	charge density	C/m^3
i	electric current	A=C/s
J	current density	A/m^3
σ	electrical conductivity	S/m
η_e	electrical resistivity	Ω.m
μ	permeability	H/m
ε	absolute permittivity	F/m
η_v	dynamic or absolute viscosity	Pa.s
ν_k	kinematic viscosity (η_v/ρ)	m^2/s
ν_t	turbulent or eddy viscosity	m^2/s
κ or ζ	volume (bulk or dilatational) viscosity	Pa.s
k	thermal conductivity	W/m/K
n_e	electron density	$electrons/m^3$
n_i	ion density	$ions/m^3$
τ_e	energy confinement time	s
E_f	energy emitted per fusion reaction	J
V	volume	m^3
σ_f	fusion cross-section	barn
v_i	relative velocity of ions	m/s
χ	Deflection angle	-
ε_β	small gyroradius expansion parameter	-

Electromagnetic operators:

Parallel operator	$\nabla_{\parallel} = \frac{\vec{B}}{B^2}\vec{B}\cdot\nabla$
Perpendicular operator	$\nabla_{\perp} = \nabla - \nabla_{\parallel}$

Fluid operators:

Convection	$\frac{\partial}{\partial t} + \vec{u}\cdot\nabla$
Advection	$(\vec{v}\cdot\nabla)\vec{v}$
Vorticity	$\omega = \nabla \times \vec{v}$

Quantic operators:

Momentum	$\hat{p} = -i\hbar\nabla$
Boltzmann collision operator	$\hat{C}(f)$

Definitions

			Units (SI)
Electromagnetic	Magnetic vector potential	$\vec{A}$	V.s/m
	Magnetic field	$\vec{B} = \nabla \times \vec{A}$	T=N/m/A
	Magnetic field intensity/strength	$\vec{H} = \vec{B} / \mu$	A/m
	Electric field	$\vec{E} = -\nabla\Phi - \frac{\partial \vec{A}}{\partial t}$	V/m
	Electric displacement field	$\vec{D} = \varepsilon \cdot \vec{E}$	C/m^2
	Poynting vector	$\vec{S} = \vec{E} \times \vec{H}$	W/m^2
	Particle velocity	$\vec{v}_e$	m/s
	Electric current density	$\vec{j} = \rho_e \vec{v}_e$	A/m^2
Fluid	Flow velocity	$\vec{v}$	m/s
	Stress tensor	$\vec{s}$	N
	Shear tensor	$\vec{\tau}$	N
	Coulomb collisional frictional force density	$\langle \vec{F}_{\parallel} \rangle = -\upsilon_e m_e n_e \vec{v}_{\parallel}$	N
Energy	Hamiltonian	$\hat{H} = \hat{T} + \hat{V}$	J
	Kinetic energy	$\hat{T} = \frac{\hat{p} \cdot \hat{p}}{2m} = \frac{\hat{p}^2}{2m}$ $= -\frac{\hbar}{2m}\nabla^2$	J
	Potential energy	$\hat{V}$	J
	Gravitational potential field	$\vec{\Psi}$	J/kg
	Gravitational acceleration	$\vec{g} = -\nabla\psi$	m/s^2

Velocity	Particle velocity	$v=(2w_k/m)^{1/2}$, where w_k is the kinetic energy	m/s
	Local shear velocity	$v_s=\frac{\partial u}{\partial y}$	m/s
	Particle drift velocity	$v_{drift}=\frac{\vec{E}\times\vec{B}}{B^2}$	m/s
	Diamagnetic velocity	$v_{dia}=-\frac{1}{en_i}\frac{\nabla p\times\vec{B}}{B^2}$	m/s
	Directional velocity (parallel and perpendicular)	$v_\parallel \quad v_\perp$	m/s
	Effective Doppler shift velocity	$v_\nabla=\frac{mc}{eB^3}\left(\frac{1}{2}v_\perp^2+v_\parallel^2\right)(B\times\nabla B)$	m/s
	Pitch angle	$\alpha=\tan^{-1}(v_\perp/v_\parallel)$	-
	Magnetic or grad-B drift	$\frac{\mu}{m\omega}\vec{b}\times\nabla B$	m/s
	Inertial drift	$\frac{v_\parallel}{\omega}\vec{b}\times\frac{d\vec{b}}{dt}$	m/s
	Polarization drift	$\frac{\vec{b}}{\omega}\times\frac{dv_E}{dt}$	m/s
	Curvature drift	$\frac{v_\parallel^2}{\omega}\vec{b}\times\left(\frac{\partial\vec{b}}{\partial t}+\vec{v}_E\cdot\nabla\vec{b}\right)$	m/s
Flow	Total flow	$\vec{V}=V_\parallel+\varepsilon V_\wedge+\varepsilon^2V_\perp$	m^3/s
	Parallel flow	$\vec{V}_\parallel=\frac{(\vec{B}\cdot\vec{V})\vec{B}}{B^2}$	m^3/s

	Perpendicular flow	$\vec{V}_{\perp} = \vec{V}_{p} + \vec{V}_{\eta} + \vec{V}_{\pi}$	m^3/s
	First order perpendicular flow	$\vec{V}_{\wedge} = \vec{V}_{E} + \vec{V}_{*}$	m^3/s
	Electro-diamagnetic flow = particle drift velocity	$\vec{V}_{E} = \frac{\vec{E} \times \vec{B}}{B^2}$	m^3/s
	Diamagnetic flow	$\vec{V}_{*} = \frac{\vec{B} \times \nabla p}{nqB^2}$	m^3/s
	Polarization flow	$\vec{V}_{p} = \frac{\vec{B} \times mnd\vec{V} / dt}{nqB^2}$	m^3/s
	Frictional-force-induced flow	$\vec{V}_{\eta} = \frac{\vec{R} \times \vec{B}}{nqB^2}$	m^3/s
	Viscous-stress-induced flow	$\vec{V}_{\pi} = \frac{\vec{B} \times \nabla \cdot \vec{\pi}}{nqB^2}$	m^3/s
	Shear flow	$S = \frac{1}{2}\left(\nabla \vec{v} + (\nabla \vec{v})^{T}\right)$	m^3/s
	Pure shear flow	$C = \frac{1}{3}(\nabla \cdot \vec{v})I$	m^3/s
	Compression flow	$S_0 = S - C$	m^3/s
	Diffusion flow (Fick's law)	$\Gamma_{diff} = -D\nabla n$	m^3/s

Electromagnetic equations

			Integral	Differential
Maxwell's equations	1	Faraday-Lenz's law	$\oint_c \vec{E}\cdot dl = -\int_s \frac{\partial \vec{B}}{\partial t}\cdot ds$	$\nabla\times\vec{E} = -\frac{\partial \vec{B}}{\partial t}$
	2	Ampère-Maxwell's law	$\oint_c \vec{H}\cdot dl = \int_s \left(\frac{\partial \vec{D}}{\partial t} + \vec{j}\right)\cdot ds$	$\nabla\times\vec{H} = \frac{\partial \vec{D}}{\partial t} + \vec{j}$ $\nabla\times\vec{B} = \mu_0\left(\varepsilon_0 \frac{\partial \vec{E}}{\partial t} + \vec{j}\right)$
	3	Gauss' law	$\oint_s \vec{D}\cdot ds = Q_{free}$	$\nabla\cdot\vec{D} = q_{free}$ $\nabla\vec{E} = \frac{\rho}{\varepsilon_0}$
	4	Magnetic flux conservation	$\oint_s \vec{B}\cdot ds = 0$	$\nabla\cdot\vec{B} = 0$
Vlasov equations	1	Electron flux conservation	-	$\frac{\partial f_e}{\partial t} + v_e\cdot\nabla f_e - \frac{q}{m}\left(\vec{E} + \frac{\vec{v}_e}{c}\times\vec{B}\right)\cdot\frac{\partial f_e}{\partial v} = C_{e\to i}$
	2	Ion flux conservation	-	$\frac{\partial f_i}{\partial t} + v_i\cdot\nabla f_i - Z_i\frac{q}{m}\left(\vec{E} + \frac{\vec{v}_i}{c}\times\vec{B}\right)\cdot\frac{\partial f_i}{\partial v} = C_{i\to e}$
	-	Ohm's law	-	$\vec{E} + \vec{v}_e\times\vec{B} = \eta_e\,\vec{j}$
	-	Generalised Ohm`s law	-	$\vec{E} + \vec{v}\times\vec{B} = \eta_e\,\vec{j} + \left(\frac{1}{n_e}\right)\left(\vec{j}\times\vec{B}\right) - \frac{1}{n_e}\nabla\cdot P_e + \left(\frac{m_e}{n_e^2}\right)\frac{\partial \vec{j}}{\partial t}$ [Electric field in moving plasma]= [Resistive term]+[Hall term]+ [Electron anisotropy]+[electron inertia]

	-	Charge conservation	$\oint_s \vec{j} \cdot d\vec{s} + \frac{d}{dt}\int_V q dV = 0$	$\frac{dq}{dt} + \nabla \cdot \vec{j} = 0$
	-	Poisson's law	-	$\nabla \cdot \vec{E} = -\nabla^2 \Phi = \frac{q}{\varepsilon}$
	-	Poynting's theorem	-	$\frac{\partial w}{\partial t} = -\nabla \cdot \left(\vec{E} \times \vec{H}\right) - J \cdot \vec{E}$
	-	Particle current	$\vec{j} = \sum_j \rho_j \int \vec{v} f_j(\vec{v}, \vec{r}, t) d\vec{v}$	
	-	Particle density	$\rho_e = \sum_j \rho_j \int f_j(\vec{v}, \vec{r}, t) d\vec{v}$	
	-	Particle distribution function	$f_j(\vec{v}, \vec{r}, t) = \sum_{i=1}^{N} \delta[\vec{r} - \vec{r}_i(t)]\delta[\vec{v} - \vec{v}_i(t)]$ $N = \int_r \int_v f_j(\vec{v}, \vec{r}, t) d\vec{r} d\vec{v}$	

Fluid conservation equations

Mass conservation	$\frac{\partial \rho}{\partial t}+\nabla\cdot(\rho\vec{v})=0$
Momentum conservation (parallel + perpendicular)	$\frac{\partial \rho\vec{v}}{\partial t}+\nabla\vec{T}=0$ $\frac{\partial \rho\vec{v}}{\partial t}+\rho(\vec{v}\cdot\nabla)\vec{v}+\nabla p+\rho\nabla\vec{\Psi}+\vec{\Gamma}_{ij}=q\vec{E}+\vec{j}\times\vec{B}$ $m\frac{\partial n\vec{v}}{\partial t}+\nabla\cdot\left(p\vec{I}+\vec{\pi}+mn\vec{v}\vec{v}\right)-nq\left[\vec{E}+\vec{v}\times\vec{B}\right]-\vec{R}=0$ Cauchy momentum equation $\frac{\partial \vec{v}}{\partial t}+(\vec{v}\cdot\nabla)\vec{v}=\frac{1}{\rho}\nabla\cdot\vec{\pi}+\vec{\psi}$
Energy conservation	$\frac{\partial w}{\partial t}+\nabla\cdot\vec{w}=0$ $\frac{\partial}{\partial t}\left[\rho\frac{1}{2}mv^2\right]+\nabla\cdot\left[\rho\frac{1}{2}m\langle v^2\vec{v}\rangle-\rho q\langle\vec{E}\cdot\vec{v}\rangle\right]-\frac{m}{2}\int_v v^2\left(\frac{\partial f}{\partial t}\right)_{coll}dv=0$
Entropy conservation	$\frac{\partial s}{\partial t}+\nabla s=\frac{1}{T}\frac{\partial Q}{\partial t}$
State equation	$\frac{d}{dt}\left(p\rho^{-\gamma}\right)=0$

- Tensors:

$$A=\begin{pmatrix} a_{xx} & a_{xy} & a_{xz} \\ a_{yx} & a_{yy} & a_{yz} \\ a_{zx} & a_{zy} & a_{zz} \end{pmatrix}$$

Stress tensor	$\vec{\pi}=\vec{\pi}_{\parallel}+\varepsilon_\beta\vec{\pi}_{\wedge}+\varepsilon_\beta^2\vec{\pi}_{\perp}$
Stress tensor	$\vec{T}=\rho\,\vec{v}\vec{v}+\left(p+\frac{B^2}{2\mu_0}\right)\vec{j}-\frac{\vec{B}\vec{B}}{\mu_0}$
Maxwell stress tensor	$\sigma_{ij}=\varepsilon_0\left(E_iE_j-\frac{1}{2}\delta_{ij}E^2\right)+\frac{1}{\mu_0}\left(B_iB_j-\frac{1}{2}\delta_{ij}B^2\right)$
Collision tensor	$\vec{\Gamma}=-mn\langle v\rangle_{eff}\vec{v}$
Pressure-viscosity tensor (N/m²)	$\vec{P}=-p\vec{I}+\vec{\tau}_{\alpha\beta}$
Artificial viscosity tensor (shear tensor)	$\tau_{\alpha\beta}=-\eta_v\left(\frac{\partial v_\alpha}{\partial r_\beta}+\frac{\partial v_\beta}{\partial r_\alpha}-\frac{2}{3}\left(1-\frac{\kappa}{\eta_v}\right)\nabla\cdot\vec{v}\,\delta_{\alpha\beta}\right)$

Corresponding author: mirapas@hotmail.com

Newton's law of viscosity	$\vec{\tau} = \eta_v \left[\nabla\vec{v} + (\nabla\vec{v})^T\right] - \left(\frac{2}{3}\eta_v - \kappa\right)(\nabla\cdot\vec{v})\delta$
Navier-Stokes equation	$\rho\frac{D\vec{v}}{Dt} = -\nabla P + \nabla\cdot\left[\eta_v\left(\nabla\vec{v} + (\nabla\vec{v})^T - \frac{2}{3}(\nabla\cdot\vec{v})\vec{I}\right) + \nabla\cdot\left[\zeta(\nabla\cdot\vec{v})\vec{I}\right]\right] + \rho\vec{g}$
Euler equation	$\rho\frac{\partial\vec{v}}{\partial t} + \rho(\vec{v}\cdot\nabla)\vec{v} + \nabla p - \vec{f} = 0$
Boltzmann equation	$\frac{\partial f}{\partial t} + v\cdot\nabla f = \hat{C}$
Convection-diffusi on equation	$\frac{\partial n}{\partial t} = \nabla\cdot(D\nabla n) - \nabla\cdot(n\vec{v}) + S$

- Force:

Magnetic field force (Lorenz force):	$\vec{j}\times\vec{B}$
Electric field force:	$q\vec{E}$
Magnetic field tension force:	$\tau_m = \frac{1}{\mu_0}(\vec{B}\cdot\nabla)\vec{B}$
Shear stress	$\vec{\tau} = \mu\vec{v}_s$

- Gravitational equations

		Integral	Differential
	Gauss' law	$\oint_S \vec{g}\cdot d\vec{s} = \int_V \nabla\cdot\vec{g}\, dV$	$\nabla\cdot\vec{g} = -\nabla^2\Phi = -4\pi G\rho$

Annex F: Plasma parameters

Pressure	Magnetic field pressure	$p_m = \frac{B^2}{2\mu_0}$
	Plasma pressure (ideal gas)	$p = nk_bT$
	Pressure under anisotropic temperature hypothesis:	$p = (p_\parallel + 2p_\perp)/3$ $\pi = 2(p_\parallel - p_\perp)/3$
	Beta limit (Thermal pressure to magnetic pressure ratio)	$\beta = 2\mu_0 p / B_t^2$ $\beta = 2\mu_0 nk_bT / B_t^2$
Energy	Plasma energy	$w_p = 3nk_bT$
	Plasma entropy	$s = n\ln\left(\frac{T^{3/2}}{n}\right)$
	Total energy	$w = \left[\rho\frac{v^2}{2} + \frac{p}{\gamma-1} + \rho\psi\right] + \left[\frac{B^2}{2\mu_0}\right] + Q_{core} + Q_{rad}$
	Energy flow	$\vec{w} = \left[\left(\rho\frac{v^2}{2} + \frac{p}{\gamma-1}\right)\vec{v} + \rho\vec{\psi}\right] + \left[\frac{\vec{E}\times\vec{B}}{4\pi/c}\right]$
	Fourier heat flux density	$\vec{q} = -k^m\nabla T$ $\vec{q} = \vec{q}_\parallel + \varepsilon\vec{q}_\wedge + \varepsilon^2\vec{q}_\perp$
	Convective loss of energy:	$\frac{d}{dx}\left(\frac{1}{2}mnV_i^3 + \frac{3}{2}p_{i,\parallel}V_i\right) + \frac{d}{dy}(p_{i,\perp}V_i)$
	Debye-Hückel length in a plasma	$\lambda_D = \left(\frac{\varepsilon_r\varepsilon_0 k_b / e^2}{n_e/T_e + \sum_{ij} j^2 n_{ij}/T_i}\right)^{1/2}$ If ion mobility is negligible and ε_r=1, $\lambda_D = \left(\frac{\varepsilon_0 k_b T_e}{n_e e^2}\right)^{1/2}$
	Plasma parameter	$\Lambda = 4\pi n\lambda_D^3$
	Debye number (Debye volume)	$N_D = \frac{1}{3}\Lambda = \frac{4\pi}{3}\lambda_D^3$
	Number of particles in a Debye sphere	$\left(\frac{4\pi}{3}\lambda_D^3\right)n_e$
	Wigner-Seitz radius	$\langle r_{ws}\rangle = \left(\frac{3}{4\pi n}\right)^{1/3} = \left(\frac{3m}{4\pi Z\rho N_A}\right)^{1/3}$
	Magnetic field energy to ion rest ratio	$\beta_{mi} = \frac{B^2}{8\pi n_i m_i c^2}$

	Lundquist number (magnetic Reynolds number)	$\frac{Lv_A}{\eta_m}$
	Greenwald limit	$n_G[10^{20}m^{-3}] = \frac{I_p[10^6 A]}{\pi a^2[m]}$
	Hugill number	$H = \frac{nq_a R}{B_T}$
	Murakami number	$M = \frac{n_e R}{B}$
	Safety factor	$q_a = \frac{aB_T}{RB_p} = \frac{2\pi a^2 B_T}{\mu_0 IR}$
	Troyon limit	$\beta_{\max} = \frac{\beta_N I}{aB_0}$ with β_N=0.028
	Bohm diffusion coefficient (cm^2s^{-1})	$D_B = ck_b T/16eB$
	Coupling parameter	$\Gamma = \frac{w_c}{w_t} = \frac{e^2}{4\pi\varepsilon_0 \langle r_{ws} \rangle k_b T_e}$
	Spitzer resistivity (transverse)	$\eta_\perp$
	Spitzer resistivity (parallel)	$\eta_{\|\|}$

Annex G: Plasma particle parameters

		Electron	Ion
Frequencies	Gyrofrequency (Hz)	$\omega_{ce} = \frac{1}{2\pi}\frac{eB}{m_e c}$	$\omega_{ci} = \frac{1}{2\pi}\frac{ZeB}{m_i c}$
	Plasma frequency (Hz)	$\omega_{pe} = \frac{1}{2\pi}\left(\frac{n_e e^2}{\varepsilon_0 m_e}\right)^{1/2}$	$\omega_{pi} = \frac{1}{2\pi}\left(\frac{n_i Z^2 e^2}{\varepsilon_0 m_i}\right)^{1/2}$
	Trapping rate	$\upsilon_{Te} = \left(\lambda_e \frac{eE}{m_e}\right)^{1/2}$	$\upsilon_{Ti} = \left(\lambda_i \frac{ZeE}{m_i}\right)^{1/2}$
	Collision rate in completely ionized plasmas	$\upsilon_e = \frac{\sqrt{2T/m_e}}{\lambda}$	$\upsilon_i = \frac{\sqrt{2T/m_i}}{\lambda}$
Lengths	Wavelength	$\lambda_e = \frac{v_e}{\omega_{ce}}$	$\lambda_i = \frac{v_i}{\omega_{ci}}$
	Plasma Wavelength	$\lambda_{pe} = \frac{v_e}{\omega_{pe}}$	$\lambda_{pi} = \frac{v_i}{\omega_{pi}}$
	Electron-thermal de Broglie wavelength	$\lambda_{th,e} = \frac{h}{m_e v_e}\sqrt{1-(v_e/c)^2} = = h\sqrt{\frac{1}{2m_e k_b T_e} - \frac{1}{m_e^2 c^2}}$	-
	Classical Radius	$r_0 = \frac{e^2}{c^2 m_e}$	-
	Gyroradius	$\tau_e = \frac{v_{Te}}{\omega_{ce}}$	$\tau_i = \frac{v_{Ti}}{\omega_{ci}}$
	Internal characteristic length	$\lambda = \frac{c}{\omega_{pe}} = \sqrt{c^2 m_e \varepsilon_0 / ne^2}$	-
	Plasma skin depth / ion inertial length	$d_e = \frac{c}{\omega_{pe}}$	$d_i = \frac{c}{\omega_{pi}}$
	Classical distance of closest approach	$r_c = \frac{e^2}{4\pi\varepsilon_0 T}$	-
	Mean free path	$\lambda_e = \frac{\overline{v_e}}{\upsilon_e}$	$\lambda_i = \frac{\overline{v_i}}{\upsilon_i}$
Velocities	Thermal velocity (most probable)	$v_{Te} = \left(\frac{2k_b T_e}{m_e}\right)^{1/2}$	$v_{Ti} = \left(\frac{2k_b T_i}{m_i}\right)^{1/2}$
	Thermal velocity (total in 3D)	$v_{Te} = \left(\frac{3k_b T_e}{m_e}\right)^{1/2}$	$v_{Ti} = \left(\frac{3k_b T_i}{m_i}\right)^{1/2}$

	Thermal velocity (mean atoms/molecules)	$v_{Te} = \left(\frac{8k_bT_e}{m_e\pi}\right)^{1/2}$	$v_{Ti} = \left(\frac{8k_bT_i}{m_i\pi}\right)^{1/2}$
	Alfvén velocity	-	$v_A = \frac{B}{(\mu_0 n_i m_i)^{1/2}}$
	Ion speed of sound	-	$c_s = \left(\frac{\gamma_e Z k_b T_e + \gamma_i k_b T_i}{m_i}\right)^{1/2}$
	Particle speed to speed of light ratio	$\beta_e = v_e/c$	$\beta_i = v_i/c$
	Lorentz factor	$\gamma_e = \sqrt{1-\beta_e^2}$	$\gamma_i = \sqrt{1-\beta_i^2}$
Force	Collisional moment	$\vec{R}_e \cong -m_e n_e \upsilon_e (\vec{v}_e - \vec{v}_i)$	$\vec{R}_i = -\vec{R}_e$
Electromagnetic	Particle magnetic moment	$\mu_e = \frac{e}{2c}\vec{r}\times\vec{v}_e$	$\mu_i = \frac{Ze}{2c}\vec{r}\times\vec{v}_i$
Temperature	Kinetic temperature	T_e	$T_i = \frac{1}{3}m_i\langle v^2\rangle$
Density	Particle density (Maxwell-Boltzmann law)	$n_e = \frac{n_{e0}}{G_e} g_j e^{-\varepsilon_j/kT}$	$n_i = \frac{n_{i0}}{G_i} g_j e^{-\varepsilon_j/kT_i}$
	Partition function	$G_e = \sum_j g_j e^{-\varepsilon_j/kT_e}$	$G_i = \sum_j g_j e^{-\varepsilon_j/kT_i}$
Energy	Energy	$w_e = \frac{1}{2}m_e v_e^2 + q_e\phi$	$w_i = \frac{1}{2}m_i v_i^2 + q_i\phi$
	Thermal energy per particle	$E_t = k_b T_e$	$E_t = k_b T_i$
	Coulomb energy per particle	$w_{ce} = \frac{e^2}{4\pi\varepsilon_0\langle r_{ws}\rangle}$	$w_{ci} = \frac{Z^2e^2}{4\pi\varepsilon_0\langle r_{ws}\rangle}$
	Heat flux (W)	$q_e = -k_\parallel^e \nabla T_e - 0.71\frac{T_e j}{e}$	$q_i = -k_\parallel^i \nabla T_i$

Quasi-neutrality condition (charge separation) : $\sum_{n_i} q_i + \sum_{n_e} q_e = 0 \quad n_i \cong n_e$

Relativistic plasma: $v_e > 0.86c \quad \gamma = 2 \quad T_e > 260keV$

Annex H: Fusion reaction parameters

		Units (SI)
Reaction rate	$\frac{dN}{dt} = n_a n_b \int_V \int_0^\infty \sigma(v_i)\phi(v_i)dv_i dV$	s^{-1}
Reactivity	$\langle \sigma v_i \rangle$	m^3s^{-1}
Averaged reaction rate	$\left\langle \frac{dN}{dt} \right\rangle = n_a n_b \langle \sigma v_i \rangle V$	s^{-1}
Fusion power density	$P_f = \left\langle \frac{dN}{dt} \right\rangle \frac{E_f}{V} = n_a n_b \langle \sigma_f v_i \rangle E_f$	$J/m^3/s$
Stored energy density	$W_p = \left[\frac{3}{2} n_a T_a + \frac{3}{2} n_b T_b \right] V \approx 3 n_i T_i V$	J/m^3
Power loss density	$P_{loss} = -\frac{W_p}{\tau_E} + P_{heat}$	$J/m^3/s$
Energy Gain Factor	$Q = \frac{P_f}{P_{heat}}$	-
Fusion Triple Product (FTP) or Fusion Performance Parameter (FPP) - Lawson criterion	$FTP = n_i T_i \tau_E$	$x10^{20}$ m^{-3}.keV.s
Fusion multiplication constant (1)	$k_\infty = \frac{n_{i+1}}{n_i}$	-

(1) *n = fusionable nuclide density in generation i+1 by nuclide density in generation i*

Table of contents

Printed by Books on Demand GmbH, Norderstedt / Germany